Quasianalytic Monogenic Solutions of a Cohomological Equation

[redacted]

[redacted]

MEMOIRS
of the
American Mathematical Society

Number 780

Quasianalytic Monogenic Solutions of a Cohomological Equation

S. Marmi
D. Sauzin

July 2003 • Volume 164 • Number 780 (second of 5 numbers) • ISSN 0065-9266

American Mathematical Society
Providence, Rhode Island

2000 *Mathematics Subject Classification.*
Primary 37F50, 30D60, 30G30, 11A55, 11J06, 40G10, 34M37.

Library of Congress Cataloging-in-Publication Data

Marmi, S. (Stefano), 1963–
Quasianalytic monogenic solutions of a cohomological equation / S. Marmi, D. Sauzin.
p. cm. — (Memoirs of the American Mathematical Society, ISSN 0065-9266 ; no. 780)
"Volume 164, number 780 (second of 5 numbers)."
Includes bibliographical references.
ISBN 0-8218-3325-1 (alk. paper)
1. Monogenic functions. 2. Small divisors. 3. Quasianalytic functions. 4. Continued fractions. I. Sauzin, D., 1966– II. Title. III. Series.

QA3.A57 no. 780
[QA331]
510 s–dc21
[515′.352] 2003048024

Memoirs of the American Mathematical Society

This journal is devoted entirely to research in pure and applied mathematics.

Subscription information. The 2003 subscription begins with volume 161 and consists of six mailings, each containing one or more numbers. Subscription prices for 2003 are $555 list, $444 institutional member. A late charge of 10% of the subscription price will be imposed on orders received from nonmembers after January 1 of the subscription year. Subscribers outside the United States and India must pay a postage surcharge of $31; subscribers in India must pay a postage surcharge of $43. Expedited delivery to destinations in North America $35; elsewhere $130. Each number may be ordered separately; *please specify number* when ordering an individual number. For prices and titles of recently released numbers, see the New Publications sections of the *Notices of the American Mathematical Society.*

Back number information. For back issues see the *AMS Catalog of Publications.*

Subscriptions and orders should be addressed to the American Mathematical Society, P. O. Box 845904, Boston, MA 02284-5904, USA. *All orders must be accompanied by payment.* Other correspondence should be addressed to 201 Charles Street, Providence, RI 02904-2294, USA.

Memoirs of the American Mathematical Society is published bimonthly (each volume consisting usually of more than one number) by the American Mathematical Society at 201 Charles Street, Providence, RI 02904-2294, USA. Periodicals postage paid at Providence, RI. Postmaster: Send address changes to Memoirs, American Mathematical Society, 201 Charles Street, Providence, RI 02904-2294, USA.

This publication is indexed in *Science Citation Index®*, *SciSearch®*, *Research Alert®*, *CompuMath Citation Index®*, *Current Contents®/Physical, Chemical & Earth Sciences.*
Printed in the United States of America.

∞ The paper used in this book is acid-free and falls within the guidelines established to ensure permanence and durability.
Visit the AMS home page at `http://www.ams.org/`

10 9 8 7 6 5 4 3 2 1 08 07 06 05 04 03

Contents

Abstract

We prove that the solutions of a cohomological equation of complex dimension one and in the analytic category have a monogenic dependence on the parameter, and we investigate the question of their quasianalyticity. This cohomological equation is the standard linearized conjugacy equation for germs of holomorphic maps in a neighborhood of a fixed point. The parameter is the eigenvalue of the linear part, denoted by q.

Borel's theory of non-analytic monogenic functions has been first investigated by Arnold and Herman in the related context of the problem of linearization of analytic diffeomorphisms of the circle close to a rotation. Herman raised the question whether the solutions of the cohomological equation had a quasianalytic dependence on the parameter q. Indeed they are analytic for $q \in \mathbb{C} \setminus \mathbb{S}^1$, the unit circle $\mathbb{S}^1$ appears as a natural boundary (because of resonances, i.e. roots of unity), but the solutions are still defined at points of $\mathbb{S}^1$ which lie "far enough from resonances". We adapt to our case Herman's construction of an increasing sequence of compacts which avoid resonances and prove that the solutions of our equation belong to the associated space of monogenic functions; some general properties of these monogenic functions and particular properties of the solutions are then studied.

For instance the solutions are defined and admit asymptotic expansions at the points of $\mathbb{S}^1$ which satisfy some arithmetical condition, and the classical Carleman Theorem allows us to answer negatively to the question of quasianalyticity at these points. But resonances (roots of unity) also lead to asymptotic expansions, for which quasianalyticity is obtained as a particular case of Écalle's theory of resurgent functions. And at constant-type points, where no quasianalytic Carleman class contains the solutions, one can still recover the solutions from their asymptotic expansions and obtain a special kind of quasianalyticity.

Our results are obtained by reducing the problem, by means of Hadamard's product, to the study of a fundamental solution (which turns out to be the so-called q-logarithm or "quantum logarithm"). We deduce as a corollary of our work the proof of a conjecture of Gammel on the monogenic and quasianalytic properties of a certain number-theoretical Borel-Wolff-Denjoy series.

2000 *Mathematics Subject Classification.* 37F50 30D60 30G30 11A55 11J06 40G10 34M37.

Key words and phrases. Small divisors, continued fractions, Carleman classes, Borel's monogenic functions, resurgence theory, Hadamard's product.

CHAPTER 1

Introduction

1.1. Let q a complex number, $g(z)$ a germ of holomorphic function which vanishes at 0, and consider the *one-dimensional cohomological equation*

$$f(qz) - f(z) = g(z), \tag{1.1}$$

where the unknown function f is required to vanish at 0. If $|q| \neq 1$ there is a unique solution, which can be obtained directly by iterating the equation forwards or backwards:

$$f(z) = f_g^-(q,z) = -\sum_{m\geq 0} g(q^m z), \qquad |q| < 1,$$

$$f(z) = f_g^+(q,z) = \sum_{m\geq 1} g(q^{-m} z), \qquad |q| > 1.$$

These two series are uniformly convergent in each compact subset of $\mathbb{D}\times\mathbb{D}_r$ or $\mathbb{E}\times\mathbb{D}_r$ respectively, where the factor $\mathbb{D}_r$ denotes the disk of convergence of g and the first factor corresponds to the parameter q, with

$$\mathbb{D} = \{q \in \mathbb{C} \mid |q| < 1\}, \quad \mathbb{E} = \{q \in \mathbb{C} \mid |q| > 1\}.$$

Thus we get two holomorphic functions of q and z. We will be particularly interested in their dependence on q, and specifically in the relationship between these two functions of q: Is it possible to cross the unit circle which separates one domain of analyticity from the other?

At a formal level, we obviously obtain from the Taylor expansion of $g(z) = \sum_{n=1}^{\infty} g_n z^n$ a unique power series satisfying (1.1):

$$f(z) = f_g(q,z) = \sum_{n\geq 1} g_n \frac{z^n}{q^n - 1} \tag{1.2}$$

which, as a series of functions of q and z, converges towards f_g^- in $\mathbb{D}\times\mathbb{D}_r$ and towards f_g^+ in $\mathbb{E}\times\mathbb{D}_r$. The case where $|q| = 1$ gives rise to the simplest non-trivial small divisor problem. Each root of the unity appears indeed as a "resonance", i.e. a pole for some terms of this series, and it is easy to define by an appropriate arithmetical condition a subset of full measure of $\mathbb{S}^1 = \{|q| = 1\}$ for which the series converges. Our purpose will be to investigate the behaviour of f in the neighborhood of this set but also near the roots of unity, from the point of view of regularity and asymptotic expansions.

Received by the editor February 5, 2001

1.2. Equation (1.1) arises naturally in the study of the existence of analytic conjugacies of germs of holomorphic diffeomorphisms of $(\mathbb{C}, 0)$ with their linear part $z \mapsto qz$; it is called cohomological because it is the linearization of the conjugacy equation. The study of the q-dependence is needed to investigate the dependence on parameters of Fatou components (more specifically Siegel disks) in the dynamics of families of rational maps on the Riemann sphere [Ris]. The conformal change of variables $z = e^{2\pi i w}$, $q = e^{2\pi i h}$ transforms (1.1) into

$$\mathcal{F}(w+h) - \mathcal{F}(w) = \mathcal{G}(w), \tag{1.3}$$

where the given function $\mathcal{G}(w) = g(e^{2\pi i w})$ is 1-periodic, analytic in the infinite semi-cylinder $\Im\mathrm{m}\, w > -\delta$ for some $\delta \in \mathbb{R}$ and tends to zero at infinity, and the unknown function $\mathcal{F}$ is required to have the same properties. In this form, but under the assumption that $\mathcal{G}$ be 1-periodic and analytic in the complex strip $|\Im\mathrm{m}\, w| < \delta$, the cohomological equation has been studied in detail by many authors, especially Wintner [Wi], Arnold [Ar] and Herman [He], since it is the linearization of the conjugacy equation of an analytic circle diffeomorphism to the rotation $w \mapsto w+h$. When h is real a small divisor problem occurs once again.

1.3. Let us return to the solutions of (1.1). We will call *fundamental solution* the function

$$f_\delta(q,z) = \sum_{n\geq 1} \frac{z^n}{q^n - 1}$$

which is obtained in the particular case where $g(z) = \delta(z) = \frac{z}{1-z}$. In view of (1.2), we recover the general solution f_g by using the Hadamard product with respect to z: $f_g = f_\delta \odot g$. Here, the Hadamard product of two formal series $A = \sum A_n z^n$ and $B = \sum B_n z^n$ is defined to be $A \odot B = \sum A_n B_n z^n$ (see Appendix A.1). The formula

$$F(q)g = f_g(q,\,.\,) = f_\delta(q,\,.\,) \odot g(\,.\,)$$

defines a mapping F from $\mathbb{D} \cup \mathbb{E}$ to some space of linear operators. For all $r > 0$ we denote by $H^\infty(\mathbb{D}_r)$ the Banach algebra of the functions which are holomorphic and bounded in $\mathbb{D}_r = \{|z| < r\}$ (equipped with the norm of the supremum over $\mathbb{D}_r$), and we consider the subspace $B_r = zH^\infty(\mathbb{D}_r)$ of the functions which vanish at the origin. We can consider F as a mapping

$$F = F_{r_1,r_2} \,:\, \mathbb{D} \cup \mathbb{E} \,\to\, \mathcal{L}(B_{r_1}, B_{r_2}) \tag{1.4}$$

for $r_1 > 0$ and $r_2 \in \,]0, r_1[$. This allows one to describe in a compact way all the solutions of (1.1) and to reduce most of the questions to the study of the fundamental solution.

1.4. To investigate the behaviour of the solutions for q near the unit circle, we introduce a few notations in connection with the roots of unity which appear as simple poles in (1.2). For $m \in \mathbb{N}^*$, we set $\mathcal{R}_m = \{\Lambda \in \mathbb{C} \mid \Lambda^m = 1\}$ (roots of unity of order m) and $\mathcal{R}_m^* = \{\Lambda = e^{2\pi i n/m},\ (n|m) = 1\}$ (primitive roots of order m). We will denote by

$$\mathcal{R} = \bigcup_{m\geq 1} \mathcal{R}_m = \bigsqcup_{m\geq 1} \mathcal{R}_m^*$$

the set of all roots of unity. To each $\Lambda \in \mathcal{R}$ we associate its order $m(\Lambda) = \min\{m \in \mathbb{N}^* \mid \Lambda \in \mathcal{R}_m\}$ so that $\Lambda \in \mathcal{R}^*_{m(\Lambda)}$. Considered as an analytic function in $(\mathbb{D} \cup \mathbb{E}) \times \mathbb{D}$, the fundamental solution satisfies the following easy but important identity:

$$(1.5)\quad f_\delta(q,z) = \sum_{\Lambda\in\mathcal{R}} \frac{\Lambda}{q-\Lambda}\mathcal{L}_{m(\Lambda)}(z), \quad \text{with } \forall m \geq 1,\ \mathcal{L}_m(z) = -\frac{1}{m}\log(1-z^m)$$

(see Appendix A.2, Lemma A2.1). This formula, which may be viewed as a "decomposition into simple elements", is in fact an example of *Borel-Wolff-Denjoy series* (see Section 2.2). By using the Hadamard product we immediately obtain an analogous formula for the general solution f_g, or more globally for the mapping F_{r_1,r_2}.

Such a formula suggests an analogy with meromorphic functions. Indeed, for each $\Lambda \in \mathcal{R}$, we will see that $(q-\Lambda)f_\delta(q,z)$ tends to $\Lambda\mathcal{L}_{m(\Lambda)}$ as q tends to Λ non-tangentially with respect to the unit circle (uniformly in z), i.e. f_δ behaves as a function with a simple pole at Λ. There is even a "Laurent series" at Λ: an asymptotic expansion which is valid near Λ, inside or outside the unit circle. But this asymptotic series must be divergent, since there are singularities infinitely close to Λ: the unit circle is a natural boundary of analyticity for $f_\delta(\,.\,,z)$, and the same is true for F_{r_1,r_2}.

1.5. On the other hand, we already mentioned that f_δ or F_{r_1,r_2} are defined when q lies in a special subset of $\mathbb{S}^1$. There too, restricting ourselves to Diophantine points, we will find asymptotic expansions. We will study the Gevrey properties of those various series, and discuss the question of *quasianalyticity* in the sense of Hadamard at the corresponding base-points: we say that a space $\mathcal{F}$ of functions is quasianalytic at a point q_0 if all its members admit an asymptotic expansion at q_0 and if any two functions in $\mathcal{F}$ with the same asymptotic expansion at q_0 coincide (i.e. the functions of $\mathcal{F}$ are determined by their asymptotics at q_0). The question of quasianalyticity is a classical one for the *Carleman classes*, but other spaces of functions are conceivable.

We wish also to investigate the regularity of f_δ or F_{r_1,r_2} in closed sets which intersect the unit circle. This naturally leads to study *monogenic* functions in domains which avoid the roots of unity: in spite of the natural boundary $\{|q|=1\}$, we try to connect the function in $\mathbb{D}$ and the function in $\mathbb{E}$ by some monogenic continuation which would replace analytic continuation.

Notice that, when we say that we wish to connect these two functions, our concern is not a relationship like $f_g^-(q,z) + f_g^+(q^{-1},z) = -g(z)$ (easy consequence of the definition of $f_g^\pm$) which is not "local" with respect to q.

1.6. Section 2 deals with the definition and properties of monogenic functions; it gives a framework in which the solutions of the cohomological equations fall, as shown in Section 2.4.

Section 3 is concerned with asymptotic expansions at those points of the unit circle which satisfy Diophantine inequalities. The question of quasianalyticity is answered negatively as far as one chooses a Diophantine base-point associated to a quadratic irrational and considers only the classical Carleman classes. This is in agreement with M. Herman's comment "The (solution of the) linearized equation does not seem to belong to any quasianalytic class" [He, p. 82].

Section 4 proposes a constructive way to recover any solution from its asymptotic expansion at some particular points: roots of unity (resonances) but also constant-type points display such a quasianalyticity property. The *resurgent* structure which appears at resonances allows one to elucidate completely the local behaviour of the solutions and to pass directly from the Laurent series at a given root of the unity to the whole Borel-Wolff-Denjoy series (1.5). At constant-type points we use the Hadamard product to define a quasianalytic space which contains the solutions.

Section 5 discusses some applications and generalizations of our work.

1.7. To conclude this introduction, let us add that the fundamental solution f_δ is known as *q-logarithmic series* ([Du]) but is perhaps more popular under the name of *"quantum logarithm"*. It is also related to the Weierstrass zeta function. The identities

$$f_\delta^-(q,z) = -\sum_{n\geq 1, m\geq 0} z^n q^{nm} = \sum_{m\geq 0} \frac{zq^m}{zq^m - 1} = z\frac{\partial}{\partial z}\log\prod_{m\geq 0}(1 - zq^m) \text{ if } q\in\mathbb{D},$$

$$f_\delta^+(q,z) = \sum_{n\geq 1, m\geq 1} z^n q^{-nm} = \sum_{m\geq 1} \frac{zq^{-m}}{1 - zq^{-m}} = -z\frac{\partial}{\partial z}\log\prod_{m\geq 1}(1 - zq^{-m}) \text{ if } q\in\mathbb{E}$$

show that the fundamental solution is related to the logarithmic derivative of Jacobi's infinite product ([HL], [Tr]). For fixed $q \in \mathbb{D}\setminus\{0\}$, f_δ^- is meromorphic over $\mathbb{C}$ with respect to z, with only simple poles at $z = q^{-m}, m \geq 0$. For fixed $q \in \mathbb{E}$, f_δ^+ is meromorphic over $\mathbb{C}$ with respect to z, with only simple poles at $z = q^m, m \geq 1$. On the other hand if q lies on the unit circle and satisfies some arithmetical condition, $\{|z| = 1\}$ is a natural boundary of analyticity as one can immediately check directly using (1.1) and the fact that the r.h.s. has a pole at $z = 1$ (see [Sim] for more details).

From the relation with Jacobi's infinite product it immediately follows that the Weierstrass zeta function ζ relative to the lattice $\mathbb{Z}\oplus h\mathbb{Z}$ can be expressed in terms of f_δ^-, f_δ^+ and the corresponding Eisenstein series

$$e_2 = \sum^e_{(n,m)\in\mathbb{Z}^2\setminus\{(0,0)\}} (n + mh)^{-2},$$

where the symbol $\sum^e$ denotes *Eisenstein summation* (terms corresponding to opposite indices are grouped in order to ensure convergence; see Paragraph 4.3.3 and [We, p. 14]). Indeed, if $q = e^{2\pi i h}$ and $z = e^{2\pi i w}$,

$$\zeta(w) = \frac{1}{w} + e_2 w + \sum^e_{\omega\in\mathbb{Z}\oplus h\mathbb{Z}} \frac{1}{w+\omega} = e_2 w - \pi i + 2\pi i[f_\delta^-(q,z) + f_\delta^+(q^{-1}, z^{-1})],$$

where the last equality holds for $|q| < |z| < |q|^{-1}$ ([We, p. 21] and [La, p. 248]).

Acknowledgments. This research started with a visit of the first author to the research group "Astronomie et Systèmes Dynamiques" in 1996. It has been supported by the CNR, the CNRS, the Institut de Mécanique Céleste, and by CEE contract ERB-CHRX-CT94-0460. The authors are grateful to J.-C. Yoccoz for his interest and some useful discussions, and to C. G. Moreira for an improvement of Lemma 3.3. The results of this study have been announced in various conferences between 1998 and 2000 (Bressanone, La Rochelle, Cetraro, Aussois, Pisa, IHES and IMPA), whose organizers we wish to thank.

CHAPTER 2

Monogenic Properties of the Solutions of the Cohomological Equation

The importance of Borel's monogenic functions in parameter-dependent small divisor problems was emphasized by Kolmogorov [Ko]. In his address to the 1954 International Congress of Mathematicians (the same where he first stated the theorem on invariant tori in the analytic case) he considers parameter-dependent vector fields on the two-dimensional torus and comments: "It is possible that the dependence ... on the parameter ... is related to the class of functions of the type of monogenic Borel functions ..."

In his work [Ar] on the local linearization problem of analytic diffeomorphisms of the circle, Arnold discussed in detail this issue; he complexified the rotation number but he did not prove that the dependence of the conjugacy on it is monogenic. This point was dealt with by M. Herman [He]. Later, Risler [Ris] extended considerably some parts of Herman's work showing that the parameter-dependence is Whitney-smooth also if one assumes less restrictive arithmetical conditions (i.e. the Brjuno condition used by Yoccoz in [Y1, Y2, Y3]). However he did not investigate monogenic properties. One should also mention that Whitney smooth dependence on parameters has been established also in the more general framework of KAM theory by Pöschel [Pö] who did not however consider complex frequencies.

Borel [Bo] wanted to extend the notion of holomorphic function so as to allow, in certain situations, analytic continuation through what is considered as a natural boundary of analyticity in Weierstrass' theory. One of his goals was apparently to determine, with the help of Cauchy's formula, not too restrictive conditions which would have ensured uniqueness of the continuation, i.e. a quasianalyticity property (see [Th]).

Extending the presentation given in [He, III.16], we recall in Section 2.1 some properties of $\mathcal{C}^1$ (and $\mathcal{C}^\infty$)-holomorphic mappings on a compact subset K of $\mathbb{C}$ with values in an arbitrary complex Banach space B. These are $\mathcal{C}^1$ maps in the sense of Whitney [Wh] which satisfy the Cauchy-Riemann condition. Being the uniform limits of B-valued rational functions with poles outside K, $\mathcal{C}^1$-holomorphic maps on K share many properties of holomorphic functions. In particular Cauchy's Theorem and Cauchy's Formula hold, and they are automatically $\mathcal{C}^\infty$-holomorphic on a subdomain of K.

Following Borel's memoir [Bo], we define in Section 2.2 the space of B-valued monogenic functions associated to an increasing sequence of compact subsets of $\mathbb{C}$ as the projective limit of the corresponding sequence of spaces of $\mathcal{C}^1$-holomorphic functions. Borel's quasianalyticity theorem for monogenic functions is then recalled, in a refined form extracted from [Wk].

In Section 2.3 we construct an increasing sequence K_j of compact sets whose union has a full-measure intersection with the unit circle. We prove in Section 2.4

that the map $q \mapsto F_{r_1,r_2}(q)$ belongs to the associated space of monogenic functions. This implies that there exists an increasing sequence of smaller compact sets $K^*_{A,j}$ on which our map is $\mathcal{C}^\infty$-holomorphic (Section 2.5).

Unfortunately the assumptions of Borel's quasianalyticity theorem are too restrictive to be applied to F_{r_1,r_2}. This is not too surprising since Borel's result is much more general and includes also monogenic functions with singularities which are dense in an open subset of $\mathbb{C}$. The problem of the quasianalyticity of $q \mapsto F_{r_1,r_2}(q)$ is addressed in Sections 3 and 4.

2.1. $\mathcal{C}^1$-holomorphic and $\mathcal{C}^\infty$-holomorphic functions

2.1.1. Let $(B, \| \ \|)$ be a complex Banach space. The following definition is taken from [He] and makes use of the generalization of the notion of smoothness of a function to a closed set due to Whitney ([St], [Wh]).

DEFINITION 2.1. Let C a closed subset of $\mathbb{C}$. A continuous function $f : C \to B$ is said to be $\mathcal{C}^1$-holomorphic if there exists a continuous map $f^{(1)} : C \to B$ such that

$$\forall z \in C,\ \forall \varepsilon > 0,\ \exists \delta > 0 \ / \quad \forall z_1, z_2 \in C,\ |z_1 - z| < \delta,\ |z_2 - z| < \delta \\ \Rightarrow \|f(z_2) - f(z_1) - f^{(1)}(z_1)(z_2 - z_1)\| \le \varepsilon |z_1 - z_2|.$$

Notice that $f^{(1)}$ in the above definition is a complex derivative: $\bar\partial f = 0$, $\partial f = f^{(1)}$ and f is holomorphic in the interior of C.

If C is compact then $\mathcal{C}^1_{hol}(C, B)$ will denote the Banach space obtained by taking as norm

$$|||f||| = \max\Big(\sup_{z \in C} \|f(z)\|, \ \sup_{z\in C} \|f^{(1)}(z)\|, \\ \sup_{z_1, z_2 \in C,\ z_1 \ne z_2} \frac{\|f(z_2) - f(z_1) - f^{(1)}(z_1)(z_2 - z_1)\|}{|z_1 - z_2|} \Big)$$

(see [ALG], Remark III.4 and Proposition III.8: in their terminology our functions define W-Taylorian 1-fields; see also [Gl], pp. 65–66).

Let K be a compact non-empty subset of $\mathbb{C}$ and let $\mathcal{C}(K, B)$ denote the uniform algebra of continuous B-valued functions on K. Let $\mathcal{R}(K, B)$ denote the uniform algebra of continuous functions from K to B which are uniformly approximated by rational functions with all the poles outside K. Let $\mathcal{O}(K, B)$ denote the uniform algebra of functions of $\mathcal{C}(K, B)$ which are holomorphic in the interior of K. Notice that f belongs to one of these uniform algebras if and only if $\ell \circ f$ belongs to the corresponding $\mathbb{C}$-valued algebra for all $\ell \in B^*$.

The inclusions

$$\mathcal{R}(K, B) \subset \mathcal{O}(K, B) \subset \mathcal{C}(K, B)$$

are in general proper; it is not too difficult to construct examples ("swiss cheeses") of compacts K with empty interior such that $\mathcal{R}(K, \mathbb{C}) \ne \mathcal{O}(K, B) = \mathcal{C}(K, \mathbb{C})$ (see [Ga] and the construction of monogenic functions below for more details).

PROPOSITION 2.1. $\mathcal{C}^1_{hol}(K, B) \subset \mathcal{R}(K, B)$.

PROOF. Let $f \in \mathcal{C}^1_{hol}(K,B)$. By Whitney's extension theorem ([Wh], Theorem I, see also [ALG], Theorem III.5) f admits a continuously differentiable extension F to a neighborhood of K. But according to Theorem 1.1 of [Ga], for all $\ell \in B^*$, the function $g = \ell \circ f$ which admits a continuously differentiable extension to a neighborhood of K and satisfies $\bar{\partial} g \equiv 0$ on K necessarily belongs to $\mathcal{R}(K,\mathbb{C})$. Hence $f \in \mathcal{R}(K,B)$. □

REMARK 2.1. As noticed by Herman, functions in $\mathcal{C}^1_{hol}(K,B)$ share some of the properties of holomorphic functions. Let $(U_\ell)_{\ell\geq 1}$ be the connected components of $\mathbb{C} \setminus K$ and assume that each ∂U_ℓ is a piecewise smooth Jordan curve. If $\sum_{\ell\geq 1} \operatorname{length}(\partial U_\ell) < +\infty$, Cauchy's theorem holds:

$$\sum_{\ell=1}^{\infty} \int_{\partial U_\ell} f(z)\, dz = 0.$$

Indeed, since $f \in \mathcal{R}(K,B)$, one can approximate f by a sequence $(r_k)_{k\in N}$ of B-valued rational functions with poles off K. Cauchy's theorem applies to these rational functions and one can pass to the limit since the convergence is uniform. Moreover, if $z \in K$ satisfies

$$\sum_{\ell=1}^{\infty} \int_{\partial U_\ell} \frac{|d\zeta|}{|\zeta - z|} < +\infty,$$

Cauchy's formula also holds:

$$f(z) = \frac{1}{2\pi i} \sum_{\ell=1}^{\infty} \int_{\partial U_\ell} \frac{f(\zeta)}{\zeta - z}\, d\zeta.$$

However to define higher order derivatives by means of Cauchy's formula one needs further assumptions on z (namely $\sum_{\ell=1}^{\infty} \int_{\partial U_\ell} \frac{|d\zeta|}{|\zeta - z|^{n+1}} < +\infty$ to obtain a derivative of order n).

2.1.2. The following definition, which generalizes Whitney $\mathcal{C}^\infty$-smoothness to the complex case, is taken from [Ri].

DEFINITION 2.2. Let C a closed subset of $\mathbb{C}$. A function $f : C \to B$ is said to be $\mathcal{C}^\infty$-holomorphic if there exist an infinite sequence of continuous functions $(f^{(n)})_{n\in\mathbb{N}} : C \to B$ such that $f^{(0)} = f$ and, for all $n, m \geq 0$, the function $R^{(n,m)}$ defined by

$$f^{(n)}(z_2) = \sum_{h=0}^{m} \frac{f^{(n+h)}(z_1)}{h!}(z_2 - z_1)^h + R^{(n,m)}(z_1,z_2), \qquad z_1, z_2 \in C,$$

satisfies the following property:

$$\forall z \in C,\ \forall \varepsilon > 0,\ \exists \delta > 0\ /$$
$$\forall z_1, z_2 \in C,\ |z_1 - z| < \delta,\ |z_2 - z| < \delta \ \Rightarrow\ \|R^{(n,m)}(z_1,z_2)\| \leq \varepsilon |z_1 - z_2|^m.$$

Clearly $\mathcal{C}^\infty$-holomorphic B-valued functions on a compact set form a Fréchet space. Once again the derivatives are taken in a complex sense, thus $\bar{\partial} f^{(n)} = 0$ for all $n \in \mathbb{N}$. The functions $f^{(n)}$ are some generalized "weak derivatives for f"; clearly f must be analytic in the interior of C and

$$\forall n, m \in \mathbb{N}, \quad \forall z \in \operatorname{int}(C), \quad f^{(n+m)}(z) = \partial^m f^{(n)}(z).$$

Whitney's extension theorem applies again: any $f \in \mathcal{C}^\infty_{hol}(C, B)$ admits an infinitely differentiable extension F to $\mathbb{C} \simeq \mathbb{R}^2$. Moreover for any $n \in \mathbb{N}$, $\partial^n F$ extends $f^{(n)}$, but of course F is not unique and $\bar{\partial} F$ need not vanish outside C.

2.2. Borel's monogenic functions

DEFINITION 2.3. Let B a complex Banach space and $(K_j)_{j\in\mathbb{N}}$ an increasing sequence of compact subsets of $\mathbb{C}$. The associated space of *B-valued monogenic functions* is defined to be the projective limit

$$\mathcal{M}((K_j), B) = \varprojlim \mathcal{C}^1_{hol}(K_j, B).$$

The restrictions $\mathcal{C}^1_{hol}(K_{j+1}, B) \to \mathcal{C}^1_{hol}(K_j, B)$ are continuous linear operators between Banach spaces, thus $\mathcal{M}((K_j), B)$ is a Fréchet space with seminorms $\|\cdot\|_{\mathcal{C}^1_{hol}(K_j,B)}$.

The above definition is inspired by the work of Borel [Bo] (see also [He], p. 81). Borel considered the case $B = \mathbb{C}$ and wanted to extend the notions of holomorphic function and analytic continuation. In the usual process of analytic continuation (defined by means of couples $([f], D(z_0, r))$ where $[f]$ is the germ at z_0 of a function analytic in the open disk $D(z_0, r)$), the domain of holomorphy of a function is necessarily open and one cannot distinguish between the points on a natural boundary of analyticity (see the discussion in [Re], Chapter V, for a nice elementary introduction, which is also related to Borel-Wolff-Denjoy series defined below). Borel's idea was to allow monogenic continuation through natural boundaries of analyticity[1] by selecting points at which the function is $\mathcal{C}^1$-holomorphic. If the function is moreover $\mathcal{C}^\infty$-holomorphic at such a point, the question of quasianalyticity may be raised: Is the function determined by its Taylor series? Such a uniqueness property could depend on the choice of the sequence (K_j) which defines the monogenic class (and not only on the union of the K_j's), and the Cauchy formula could help to establish it.

In the rest of Section 2.2, we illustrate the previous definition by a construction due to Borel of a special sequence (K_j) which is adapted to the case of *Borel-Wolff-Denjoy series* [Gou, Bo, Wo, De, Si]. They are the most studied examples of monogenic functions, and quasianalyticity can be proved in their case under suitable assumptions.

Let $\omega = (\omega_\nu)_{\nu\geq 1}$ a bounded sequence of points in $\mathbb{C}$ and $\Omega = \{\omega_\nu\}$. We will exclude smaller and smaller disks around these points; the open disk of center ω_ν and radius ρ will be denoted by $D(\omega_\nu, \rho)$. Let G be an open bounded Jordan domain which contains Ω. We fix a sequence $(r_\nu)_{\nu\in\mathbb{N}^*} \in \ell^1(\mathbb{R}^+)$ and define

$$K_j = \overline{G} \setminus \bigcup_{\nu\geq 1} D(\omega_\nu, 2^{-j} r_\nu), \quad C = \bigcup_{j\geq 1} K_j. \tag{2.1}$$

[1] M. Herman pointed out to us that Poincaré himself investigated the possibility of generalizing Weierstrass' process of analytic continuation so as to consider functions whose singular points are dense on an open set or a Jordan curve [P1, P2].

Notice the inclusions

$$\overline{G} \setminus \overline{\Omega} \subset C \subset \overline{G} \setminus \Omega,$$

which are in general proper.

For each each sequence $a = (a_\nu)_{\nu \geq 1} \in \ell^1(B)$, we can define a function

$$\Sigma_\omega(a) : q \mapsto \big(\Sigma_\omega(a)\big)(q) = \sum_{\nu=1}^{\infty} \frac{a_\nu}{q - \omega_\nu}$$

which is holomorphic in $\mathbb{C} \setminus \overline{\Omega}$. We get a linear operator $\Sigma_\omega : \ell^1(B) \to \mathcal{O}(\mathbb{C} \setminus \overline{\Omega}, B)$ which is generally not injective (see [Wo] for some examples). But we have also the following

LEMMA 2.1. *The operator* Σ_ω *induces an injective operator from the space*

$$\ell^1_r(B) = \{a = (a_\nu)_{\nu \geq 1} \in \ell^1(B) \mid \forall \nu \geq 1,\ \|a_\nu\|^{1/4} < r_\nu\}.$$

into $\mathcal{M}((K_j), B)$.

PROOF. Since for all $q \in K_j$ and $\nu \geq 1$, $|q - \omega_\nu| \geq 2^{-j} r_\nu \geq 2^{-j} \|a_\nu\|^{1/4}$, it is easy to check that $\Sigma_\omega(a)|_{K_j} \in \mathcal{C}^1_{hol}(K_j, B)$ for all $j \geq 1$.

To prove injectivity we make use of a residue computation. Let $f_j = \Sigma_\omega(a)|_{K_j}$. Let $\gamma_j^{(\mu)} = \partial D(\omega_\mu, 2^{-j} r_\mu)$ with positive orientation and let $\Gamma_j^{(\mu)}$ denote the curve obtained from $\gamma_j^{(\mu)}$ replacing those parts which are covered by disks $D(\omega_\nu, 2^{-j} r_\nu)$ with $\nu \neq \mu$ by the corresponding arcs of circles $\partial D(\omega_\nu, 2^{-j} r_\nu)$ which are contained in K_j. Clearly $\Gamma_j^{(\mu)}$ is a countable union of arcs of circle, all positively oriented, and the length of $\Gamma_j^{(\mu)}$ is bounded by $2^{-j} \sum_{\nu=1}^{\infty} r_\nu$. If $G_j^{(\mu)}$ denotes the domain of $\mathbb{C}$ enclosed by $\Gamma_j^{(\mu)}$,

$$\frac{1}{2\pi i} \int_{\Gamma_j^{(\mu)}} f_j(q) dq = \sum_{\omega_\nu \in G_j^{(\mu)}} a_\nu .$$

The sequence $\nu(\mu, j) = \inf\{\nu \in \mathbb{N}^* \mid \omega_\nu \in G_j^{(\mu)},\ \omega_\nu \neq \omega_\mu\}$ tends to infinity as $j \to \infty$, thus

$$\|\frac{1}{2\pi i} \int_{\Gamma_j^{(\mu)}} f_j(q) dq - a_\mu\| \leq \sum_{\nu(\mu,j)}^{\infty} \|a_\nu\| \ \to 0 \quad \text{as} \quad j \to \infty.$$

This implies injectivity. □

Of course, if none of the coefficients a_ν vanishes, $\Sigma_\omega(a)$ is not analytic at any point of C which is an accumulation point of the sequence ω. Borel's example ([Bo], p. 144) is $B = \mathbb{C}$, $\{\omega_\nu\} = \{\frac{r+si}{n};\ 1 \leq r, s \leq n,\ (r,n) = 1,\ (s,n) = 1\}$, $a_\nu = \exp(-\exp(n^4))$ and $G = \{q \in \mathbb{C} \mid 0 < \Re\mathrm{e}\, q < 1,\ 0 < \Im\mathrm{m}\, q < 1\}$.

A remarkable result of Borel and Winkler is the following (see also [Tj])

THEOREM 2.1. *We still use the notations* (2.1) *and assume furthermore that* $r_\nu < 1$ *for all* $\nu \in \mathbb{N}^*$ *and*

$$\sum_{\nu=1}^{\infty} \left(\log \frac{1}{r_\nu}\right)^{-1} < +\infty. \tag{2.2}$$

Let

$$K_j^* = G \setminus \bigcup_{\nu=1}^{\infty} D\Big(\omega_\nu, 2^{-j}\big(\log \tfrac{1}{r_\nu}\big)^{-1}\Big), \quad C^* = \bigcup_{j=1}^{\infty} K_j^*.$$

C^* is included in C and if $f \in \mathcal{M}((K_j), B)$, the restriction $f|_{K_j^*}$ is $\mathcal{C}^\infty$-holomorphic for all $j \geq 1$. Moreover, if there exist $q_0 \in C^*$ and $j \in \mathbb{N}^*$ such that
(i) there exists a straight line s such that $q_0 \in s \cap G \subset K_j^*$,
(ii) $f^{(n)}(q_0) = 0$ for all $n \geq 0$,
the function f vanishes identically on K_j^*.

In particular, according to the definition of quasianalyticity given in Section 1.5, $\mathcal{M}((K_j), B)$ is quasianalytic at all points of C^* which satisfy the condition *(i)*. We refer to [Wk] for a proof of Theorem 2.1 (in the case where $B = \mathbb{C}$, but this restriction is not essential).

REMARK 2.2. Borel (without using Whitney's extension theorem) also proves that Cauchy's formula holds: let γ a simple positively oriented closed curve bounding a simply connected region D of G. Let γ_j denote the curve obtained from γ by replacing those parts of γ which are covered by disks $D(\omega_\nu, 2^{-j} r_\nu)$ by the corresponding parts of the circles $\partial D(\omega_\nu, 2^{-j} r_\nu)$ which are contained in $K_j \cap D$ (see [Wk] and [Ar], Section 7, for more details). Let Γ_j denote the union of those parts of the circles $\partial D(\omega_\nu, 2^{-j} r_\nu)$ which are contained in $K_j \cap D$ and not part of γ_j. Then

$$f^{(n)}(q) = \frac{n!}{2\pi i}\Big(\int_{\gamma_j} \frac{f(w)}{(w-q)^{n+1}} dw - \int_{\Gamma_j} \frac{f(w)}{(w-q)^{n+1}} dw\Big), \qquad q \in K_j^* \cap D,\ n \in \mathbb{N}.$$

REMARK 2.3. The previous theorem was proved by Winkler under less restrictive assumptions than those originally required by Borel, using Carleman's Theorem (see [Ca] and Theorem 3.1 below). Note that it holds *without any further assumption on the distribution of the singular points* $(\omega_\nu)_{\nu \geq 1}$, while for the problem we are interested in roots of unity will play a role in the sequel. The quasianalyticity properties of Borel-Wolff-Denjoy series are studied also in [Be1], [Be2] and [Si] (which focus in fact on the broader question of the injectivity of Σ_ω).

REMARK 2.4. Unfortunately one cannot apply the previous theorem to the solutions of cohomological equations since the condition (2.2) is too restrictive for that situation. Let $0 < \rho_2 < \rho_1$, $B = \mathcal{L}(B_{\rho_1}, B_{\rho_2})$ and consider the mapping (1.4). Ordering the primitive roots of unity by increasing order (i.e. following the Farey ordering of rational numbers), one can write it as a Borel-Wolff-Denjoy series

$$F(q) = \Sigma_{\mathcal{R}}(a)(q) = \sum_{\nu=1}^{\infty} \frac{\Lambda_\nu}{q - \Lambda_\nu} \mathcal{L}_{m(\nu)} \odot, \qquad \mathcal{R} = \{\Lambda_1, \Lambda_2, \dots\}, \tag{2.3}$$

setting $a_\nu = \Lambda_\nu \mathcal{L}_{m(\nu)} \odot$. Since the number of terms in the Farey series of order m is approximately $\frac{3m^2}{\pi^2}$ ([HW], Theorem 331, p. 268) one has $m(\nu) \simeq \frac{\pi}{\sqrt{3}}\sqrt{\nu}$. On the other hand, one checks easily that $\|a_\nu\| \simeq \frac{1}{m(\nu)}\big(\frac{\rho_2}{\rho_1}\big)^{m(\nu)}$. The requirement $(a_\nu)_{\nu \geq 1} \in \ell_r^1(B)$ leads to a lower bound $r_\nu \geq c_1 c_2^{m(\nu)/4}$ and the condition (2.2) is violated.

2.3. Domains of monogenic regularity: The sequence (K_j)

The goal of this section is to specify a sequence of compact sets $(K_j)_{j\in\mathbb{N}}$ so as to be able to prove (in Section 2.4) that $\mathcal{M}((K_j), B)$ contains the solutions of the cohomological equation. In the definitions of the domains $C_{\psi,\kappa,d}$ and $W^A_{\gamma,\kappa,d}$ given below (Definitions 2.4 and 2.6), we will follow a construction given by M. Herman [He] for Diophantine numbers (see also [Ris] for a similar construction for Brjuno numbers). We adapt it slightly so as to deal with more general irrational numbers.

2.3.1. The conformal change of variable $q = e^{2\pi i x}$ maps $\mathbb{C}^*$ biholomorphically on $\mathbb{C}/\mathbb{Z}$, the circle $\{|q| = 1\}$ on $\mathbb{R}/\mathbb{Z}$ and $\mathcal{R}^*_m$ on $\{\frac{n}{m} \mid m \in \mathbb{N}^*,\ 0 \le n \le m-1,\ (n,m) = 1\}$. We will use the notations of Appendix A.3 for continued fractions: if $x \in \mathbb{R}\setminus\mathbb{Q}\,(\mathrm{mod}\,\mathbb{Z})$, we will denote by $[0, a_1(x), a_2(x), \dots]$ its continued fraction expansion and by $\left(\frac{n_k(x)}{m_k(x)}\right)_{k\ge 0}$ the corresponding sequence of convergents, omitting sometimes the dependence on x. Note that $n_0/m_0 = 0/1$.

DEFINITION 2.4. We call an approximation function any decreasing function ψ on $\mathbb{N}^*$ such that

$$2\sum_{m=1}^{+\infty} \psi(m) < 1 \quad \text{and} \quad \forall m \ge 1,\ 0 < \psi(m) \le \frac{1}{2m}.$$

We associate with it a subset of $\mathbb{R}\setminus\mathbb{Q}\,(\mathrm{mod}\,\mathbb{Z})$:

$$C_\psi = \Big\{x \in \mathbb{R}\setminus\mathbb{Q}\,(\mathrm{mod}\,\mathbb{Z}) \mid \forall k \ge 0,\ m_{k+1}(x) \le \frac{1}{\psi(m_k(x))}\Big\}, \tag{2.4}$$

and some subsets of $\mathbb{C}/\mathbb{Z}$ whose traces on $\mathbb{R}/\mathbb{Z}$ are C_ψ:

$$C_{\psi,\kappa} = \bigcup_{y\in C_\psi} \{x \in \mathbb{C}/\mathbb{Z} \mid |\Im\mathrm{m}\,\tilde{x}| \ge \kappa|\Re\mathrm{e}(\tilde{x}-\tilde{y})|\}, \quad C_{\psi,\kappa,d} = C_{\psi,\kappa} \cap \big\{|\Im\mathrm{m}\,x| \le d\big\},$$

for $\kappa \in]0,1[$ and $d > 0$, where $\tilde{x}$ and $\tilde{y}$ denote some lifts in $\mathbb{C}$ of x and y (see Figure 2).

Notice that C_ψ consists of points which are "far enough from the rationals", as measured by ψ; namely, according to (A3.4) and Proposition A3.2,

$$\bigcap_{n/m} \{x \mid |x - \frac{n}{m}| \ge \frac{\psi(m)}{m}\} \subset C_\psi \subset \bigcap_{n/m} \{x \mid |x - \frac{n}{m}| > \frac{\psi(m)}{2m}\}. \tag{2.5}$$

The most interesting case for our purposes will be $\psi(m) = \gamma\, e^{-\alpha m}$ with fixed $\alpha > 0$ and $\gamma \in]0, \inf(\frac{\alpha e}{2}, \frac{e^\alpha - 1}{2})[$. The classical Diophantine condition of exponent $\tau > 2$ (see Section 3.2) would correspond to $\bigcup C_{\phi_{\gamma,\tau}}$, where $\phi_{\gamma,\tau}(m) = \gamma m^{1-\tau}$ and the union is taken over those $\gamma > 0$ such that $\phi_{\gamma,\tau}$ is an approximation function (i.e. $\gamma \le \frac{1}{2\zeta(\tau-1)}$, denoting by ζ the Riemann zeta function).

The Diophantine exponent $\tau = 2$ (which is associated to constant-type points) was not considered here, only because the corresponding functions $\phi_{\gamma,2}$ do not satisfy the condition of summability in Definition 2.4. This condition is used in the next lemma to ensure positive measure for C_ψ, and indeed the set of constant-type points has measure zero.

LEMMA 2.2. *If ψ is an approximation function, C_ψ has positive Lebesgue measure. For all $\varepsilon > 0$ there exists an approximation function ψ such that*

$$|C_\psi \,(\mathrm{mod}\,\mathbb{Z})| > 1 - \varepsilon.$$

PROOF. According to (2.5), the one-dimensional Lebesgue measure of the complement $(\mathbb{R}/\mathbb{Z}) \setminus C_\psi$ is less than

$$2\sum_{m=1}^{\infty}\sum_{n=0}^{m-1}\frac{\psi(m)}{m} \le 2\sum_{m=1}^{\infty}\psi(m) < 1.$$

Given $\varepsilon > 0$, we choose $\psi(m) = \frac{\varepsilon}{2\zeta(2)m^2}$ to make the previous quantity less than ε. □

In order to investigate the structure of this kind of set, it is useful to refer to a suitable partition of $\mathbb{R}/\mathbb{Z}$ obtained by considering a finite number of iterations of the Gauss map A (see Appendix A.3 for the definition of the Gauss map; the intervals defined below are called "intervals of rank k" in [Khi]).

Let $a_1, \dots, a_k$ positive integers ($k \in \mathbb{N}^*$). We associate with them the finite continued fractions $[0, a_1, \dots, a_{k-1}] = \frac{n_{k-1}}{m_{k-1}}$ and $[0, a_1, \dots, a_k] = \frac{n_k}{m_k}$, and define an interval

$$I(a_1, \dots, a_k) = \{x = \tfrac{n_k + n_{k-1}y}{m_k + m_{k-1}y},\ y \in]0,1[\} = \begin{cases}]\frac{n_k}{m_k}, \frac{n_k+n_{k-1}}{m_k+m_{k-1}}[& \text{if } k \text{ is even} \\ ]\frac{n_k+n_{k-1}}{m_k+m_{k-1}}, \frac{n_k}{m_k}[& \text{if } k \text{ is odd} \end{cases}$$

(the alternative stems from (A3.3)). Each such interval is a branch of the k-th iterate A^k of the Gauss map, precisely the branch which is determined by the fact that all points $x \in I(a_1, \dots, a_k)$ have $\{0, a_1, \dots, a_k\}$ as first $k+1$ partial quotients (see Formula (A3.1)). For a given $k \ge 1$, the union of all branches of A^k yields a partition of $\mathbb{R}/\mathbb{Z}$

$$\forall k \ge 1, \quad \mathbb{R}/\mathbb{Z} = \mathcal{F}_k \cup \bigcup_{a_1, \dots, a_k \ge 1} I(a_1, \dots, a_k),$$

where[2] $\mathcal{F}_k = \big\{[0, a_1, \dots, a_\ell],\ 1 \le \ell \le k,\ a_i \ge 1\big\} \subset \mathbb{Q}/\mathbb{Z}$. The previous definition allows for a convenient rephrasing of (2.4):

$$C_\psi = \bigcap_{k \ge 1} \bigsqcup\nolimits_\psi I(a_1, \dots, a_k),$$

where for each $k \ge 1$, $\bigsqcup_\psi$ denotes the disjoint union over those $(a_1, \dots, a_k)$ such that $m_{i+1} \le 1/\psi(m_i)$ for $i = 0, \dots, k-1$ (here, of course, m_i is the denominator of $[0, a_1, \dots, a_i]$).

[2]Notice that any rational number $n/m \in]0,1[$ is the endpoint of exactly two branches of the iterated Gauss map. Indeed n/m can be written in a unique way as $n/m = [0, \bar{a}_1, \dots, \bar{a}_\ell]$ for some $\ell \ge 1$, with $\bar{a}_1, \dots, \bar{a}_{\ell-1} \ge 1$ and $\bar{a}_\ell \ge 2$; it is the left (right) endpoint of $I(\bar{a}_1, \dots, \bar{a}_\ell)$ and the right (left) endpoint of $I(\bar{a}_1, \dots, \bar{a}_{\ell-1}, \bar{a}_\ell - 1, 1)$ if ℓ is even (odd).

2.3.2. We will indicate some more properties of the set C_ψ associated to an approximation function ψ. As a preliminary, to each rational number $n/m \in \mathbb{Q}/\mathbb{Z}$ we attach an open interval $\mathcal{J}_\psi(n/m)$ such that

$$n/m \in \mathcal{J}_\psi(n/m) \subset (\mathbb{R}/\mathbb{Z}) \setminus C_\psi. \tag{2.6}$$

To define it we proceed as follows:

(i) if $n/m = 0/1$, we set

$$\mathcal{J}_\psi(0/1) := \mathrm{int}\Big(\bigcup_{a_1 > 1/\psi(1)} \overline{I(a_1)} \cup \bigcup_{a_2+1>1/\psi(1)} \overline{I(1,a_2)} \Big); \tag{2.7}$$

(ii) if $n/m \neq 0/1$ and $(n,m) = 1$, we write $n/m = [0, \bar{a}_1, \dots, \bar{a}_k]$, with $k \geq 1$, $\bar{a}_1, \dots, \bar{a}_{k-1} \geq 1$ and $\bar{a}_k \geq 2$, we set $n_-/m_- = [0, \bar{a}_1, \dots, \bar{a}_{k-1}]$ (if $k \geq 2$; otherwise $n_-/m_- = 0/1$) and

$$\mathcal{J}_\psi(n/m) := \mathrm{int}\Big(\bigcup_{a_{k+1}m+m_- > 1/\psi(m)} \overline{I(\bar{a}_1, \dots, \bar{a}_k, a_{k+1})} \cup \bigcup_{(a_{k+2}+1)m-m_->1/\psi(m)} \overline{I(\bar{a}_1, \dots, \bar{a}_k - 1, 1, a_{k+2})} \Big). \tag{2.8}$$

This definition is motivated by the relations (A3.0). For instance the points in the first union of (2.8) have continued fraction expansions such that $a_{k+1}m + m_- = m_{k+1}$ since $m_{k-1} = m_-$ and $m_k = m$ for them, and in the second one, $(a_{k+2}+1)m - m_- = m_{k+2}$ since $m_k = m - m_-$ and $m_{k+1} = m$ (except at one of the boundary-points of each interval: $m_{k+1} = (a_{k+1}+1)m + m_-$ and $m_{k+2} = (a_{k+2}+2)m - m_-$ respectively, for these exceptional rational points). We thus have $m_{k+1} > 1/\psi(m_k)$ or $m_{k+2} > 1/\psi(m_{k+1})$ respectively, hence $\mathcal{J}_\psi(n/m)$ is contained in the complement of C_ψ. To check that it is an open interval, consider for instance the case of odd k: using (A3.1) one can write the first union as $\{\frac{n+n_-y}{m+m_-y}\,;\ 0 < y \leq 1/M\}$, where M is the minimum value of a_{k+1} (i.e. $M = [\frac{1}{m}(\frac{1}{\psi(m)} - m_-)] + 1$), this union is thus a non-empty interval whose right endpoint is n/m; similarly the second union is a non-empty closed interval whose left endpoint is n/m. Notice that, in case (i), $\mathcal{J}_\psi(0/1) = \mathrm{int}([0, \frac{1}{M}] \cup [1 - \frac{1}{M}, 1])$ must be identified with $]-\frac{1}{M}, \frac{1}{M}[$ (where $M = [\frac{1}{\psi(1)}] + 1$).

LEMMA 2.3. *The set C_ψ associated to any approximation function is totally disconnected, closed and perfect.*

PROOF. Since $C_\psi \cap (\mathbb{Q}/\mathbb{Z}) = \emptyset$, C_ψ is totally disconnected. To see that C_ψ is closed observe that, if $x \in (\mathbb{R}/\mathbb{Z}) \setminus C_\psi$, one can exhibit an open neighborhood of x which is contained in the complement of C_ψ: either $x \in \mathbb{Q}$ and $\mathcal{J}_\psi(x)$ is such a neighborhood, or $x \notin \mathbb{Q}$ and $m_{k+1}(x) > 1/\psi(m_k(x))$ for some $k \geq 0$, hence $I(a_1(x), \dots, a_{k+1}(x))$ will do.

We now prove that any $x \in C_\psi$ is an accumulation point of C_ψ. For each $j \in \mathbb{N}^*$ we define a linear fractional map $T_{x,j}$ mapping any $y \in]0,1[$ to

$$T_{x,j}(y) = \frac{n_j(x) + n_{j-1}(x)y}{m_j(x) + m_{j-1}(x)y} = [0, a_1(x), \dots, a_j(x), a_1(y), a_2(y), \dots].$$

Let us use $y^* = \frac{\sqrt{5}-1}{2} = [0,1,1,\dots]$. The sequence $x^{(j)} = T_{x,j}(y^*)$ converges to x as $j \to \infty$, and one can check that each $x^{(j)} \in C_\psi$:

– If $k \le j$, $\frac{n_k(x^{(j)})}{m_k(x^{(j)})} = \frac{n_k(x)}{m_k(x)}$; thus $m_{k+1}(x^{(j)}) \le \frac{1}{\psi(m_k(x^{(j)}))}$ for all $k \le j-1$ and

$$m_{j+1}(x^{(j)}) = m_j(x) + m_{j-1}(x) \le a_{j+1}(x)m_j(x) + m_{j-1}(x)$$
$$= m_{j+1}(x) \le \frac{1}{\psi(m_j(x^{(j)}))}.$$

– If $k \ge j+1$, we use $\psi(m) \le 1/2m$:

$$m_{k+1}(x^{(j)}) = m_k(x^{(j)}) + m_{k-1}(x^{(j)}) \le 2m_k(x^{(j)}) \le \frac{1}{\psi(m_k(x^{(j)}))}.$$

□

The intervals $\mathcal{J}_\psi(n/m)$ defined above will also help us in the proof of the next proposition which describes the connected components of $(\mathbb{R}/\mathbb{Z}) \setminus C_\psi$.

PROPOSITION 2.2. *Let*

$$\mathbb{Q}_\psi = \{0/1\} \cup \{n/m \in \mathbb{Q}/\mathbb{Z} \mid n/m \ne 0/1 \text{ and } m_{j+1} \le 1/\psi(m_j) \text{ for } j = 0,\dots,k-1\},$$

with the usual notations and conventions: the m_j's $(0 \le j \le k)$ are the denominators of the convergents $[0,a_1,\dots,a_j]$ of $n/m = [0,a_1,\dots,a_k]$, with $a_k \ge 2$.

(1) *Each connected component of $(\mathbb{R}/\mathbb{Z}) \setminus C_\psi$ contains one and only one point of $\mathbb{Q}_\psi$, which is a convergent of both of its endpoints. We denote the connected component of $n/m \in \mathbb{Q}_\psi$ by $]x_{n/m}, x'_{n/m}[\subset \mathbb{R}/\mathbb{Z}$ (which must be identified to an open subinterval of $]0,1[$ if $n/m \ne 0/1$, or of $]-1/2,1/2[$ if $n/m = 0/1$).*

(2) $\frac{\psi(m)}{2m} \le |x - \frac{n}{m}| < \frac{2\psi(m)}{m}$ *if $x = x_{n/m}$ or $x'_{n/m}$.*

(3) *If $r/s \in]x_{n/m}, x'_{n/m}[$ and $r/s \ne n/m$, $s > \frac{1}{\psi(m)} \ge 2m$.*

PROOF. Any connected component of $U = (\mathbb{R}/\mathbb{Z}) \setminus C_\psi$ contains at least a rational r/s. Suppose this rational does not belong to $\mathbb{Q}_\psi$ and write it as $r/s = [0,a_1,\dots,a_\ell]$ with $a_\ell \ge 2$: we must have $m_{k+1} > 1/\psi(m_k)$ for some $k \ge 0$. Choosing k to be minimal, we obtain $n/m = [0,a_1,\dots,a_k] \in \mathbb{Q}_\psi$ and $r/s \in \mathcal{J}_\psi(n/m)$ (note that $n/m = 0/1$ if $k = 0$). Thus the connected component of r/s contains $\mathcal{J}_\psi(n/m)$, and in particular n/m. We notice in passing that $s > 1/\psi(m) \ge 2m$.

Let us now suppose that $]x, x'[$ is the connected component of $n/m \in \mathbb{Q}_\psi$ in U and check that n/m is a convergent of x and x'. We may suppose that $n/m \ne 0/1$. Let us write $n/m = [0,a_1,\dots,a_k]$ with $a_k \ge 2$. Denoting by m_- the denominator of $[0,a_1,\dots,a_{k-1}]$, we choose positive integers a and b such that

$$am + m_- \le 1/\psi(m), \quad (b+1)m - m_- \le 1/\psi(m)$$

(this is possible since $1/\psi(m) \ge 2m > m + m_-$). By the same kind of argument as at the end of the proof of Lemma 2.3, one can check that the points

$$x^+ = [0,a_1,\dots,a_{k-1},a_k,a,1^\infty] \text{ and } x^- = [0,a_1,\dots,a_{k-1},a_k-1,1,b,1^\infty]$$

both belong to C_ψ. But if k is even, $x^- < n/m < x^+$, and the order is reversed otherwise. Therefore $[x,x']$ is contained in $]x^-,x^+[$ (or $]x^+,x^-[$ is k is odd), and n/m

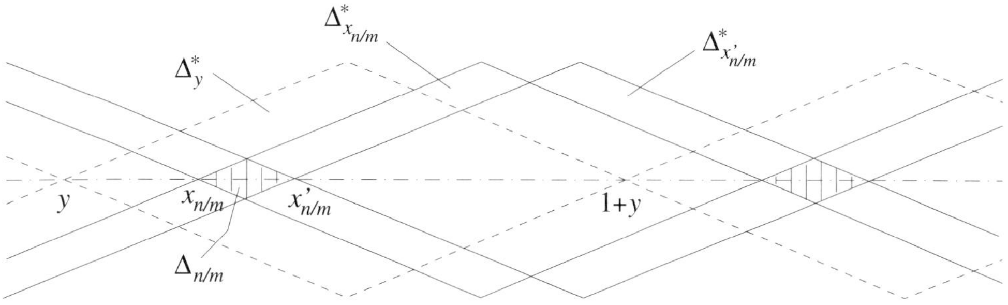

FIGURE 1. The sets $\Delta_{n/m}$ and Δ^*_y.

is a convergent of all those points. This implies easily that a connected component of U cannot contain more that one point of $\mathbb{Q}_\psi$.

The first inequality in (2) follows from the second inclusion in (2.5). For the second inequality, consider x^+ and x^- as defined above for $n/m = [0, a_1, \ldots, a_k] \in \mathbb{Q}_\psi \setminus \{0/1\}$, but this time we choose a and b maximal:

$$\frac{1}{\psi(m)} - m < am + m_- \leq \frac{1}{\psi(m)}, \quad \frac{1}{\psi(m)} - m < (b+1)m - m_- \leq \frac{1}{\psi(m)}.$$

By virtue of (A3.4), since $m_k(x^+) = m$, $m_{k+1}(x^+) = am + m_-$, $m_{k+1}(x^-) = m$ and $m_{k+2}(x^-) = (b+1)m - m_-$,

$$|x^+ - n/m|, |x^- - n/m| < \frac{\psi(m)}{m(1 - m\psi(m))} \leq \frac{2\psi(m)}{m}.$$

This yields the desired inequality. If $n/m = 0/1$, one can use $x^+ = [0, a, 1^\infty] = \frac{1}{a+g}$ with $a = [\frac{1}{\psi(1)}] \geq 2$ and $g = [0, 1^\infty]$, and $x^- = -1 + [0, 1, a-1, 1^\infty] = -x^+$.

(3) was already observed at the beginning of the proof. $\square$

2.3.3. We now fix $\kappa \in]0, 1[$ and $d > 0$, and study the sets $C_{\psi,\kappa}$ and $C_{\psi,\kappa,d}$ associated to the approximation function ψ. Proposition 2.2 yields a decomposition of $(\mathbb{R}/\mathbb{Z}) \setminus C_\psi$ into connected components; this will reflect in a description of the complement of $C_{\psi,\kappa}$:

LEMMA 2.4. *For each $n/m \in \mathbb{Q}_\psi$, let $\Delta_{n/m}$ denote the set*

$$\{x \in \mathbb{C}/\mathbb{Z} \mid \Re e\, x \in]x_{n/m}, x'_{n/m}[, \ |\Im m\, x| < \kappa \min(\Re e\, x - x_{n/m}, x'_{n/m} - \Re e\, x)\},$$

which is an open diamond of base $]x_{n/m}, x'_{n/m}[$ and slopes $\pm\kappa$ with respect to the real axis (see Figure 1). We have

$$C_{\psi,\kappa} = (\mathbb{C}/\mathbb{Z}) \setminus \bigsqcup_{n/m \in \mathbb{Q}_\psi} \Delta_{n/m}, \tag{2.9}$$

the sets $C_{\psi,\kappa,d}$ are compact subsets of $\mathbb{C}/\mathbb{Z}$ and they have positive measure when $d > \kappa\,\zeta(4)$:

$$meas\left(C_{\psi,\kappa,d}\right) > 2d - 8\kappa \sum_{m \geq 1} \left(\tfrac{\psi(m)}{m}\right)^2.$$

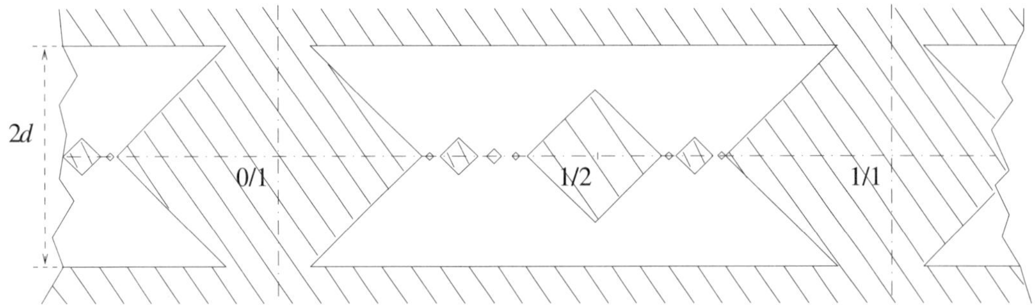

FIGURE 2. The set $C_{\psi,\kappa,d}$.

PROOF. Let us rephrase the definition of $C_{\psi,\kappa}$ as

$$C_{\psi,\kappa} = (\mathbb{C}/\mathbb{Z}) \setminus \bigcap_{y \in C_\psi} \Delta_y^*,$$

with $\Delta_y^* = \left\{x \in \mathbb{C}/\mathbb{Z} \mid |\Im\mathrm{m}\, \tilde{x}| < \kappa \min(\Re\mathrm{e}\, \tilde{x} - \Re\mathrm{e}\, \tilde{y}, 1 + \Re\mathrm{e}\, y - \Re\mathrm{e}\, \tilde{x})\right\}$ (each Δ_y^* is an open diamond whose trace on $\mathbb{R}/\mathbb{Z}$ has length 1 and coincides with the complement of $\{y\}$; see Figure 1). Formula (2.9) is now reduced to the identity

$$\bigsqcup_{n/m \in \mathbb{Q}_\psi} \Delta_{n/m} = \bigcap_{y \in C_\psi} \Delta_y^*.$$

If $n/m \in \mathbb{Q}_\psi$ and $y \in C_\psi$, the fact that $y \notin]x_{n/m}, x'_{n/m}[$ implies that $\Delta_{n/m} \subset \Delta_y^*$, hence the union of the diamonds $\Delta_{n/m}$ is contained in the intersection of the diamonds Δ_y^*. Let x in the intersection of the diamonds Δ_y^*. If $x \in \mathbb{R}/\mathbb{Z}$, this means that $x \notin C_\psi$, thus $x \in]x_{n/m}, x'_{n/m}[\subset \Delta_{n/m}$ for some $n/m \in \mathbb{Q}_\psi$. If $x \notin \mathbb{R}/\mathbb{Z}$, the intersection with $\mathbb{R}/\mathbb{Z}$ of the lines of slopes $\pm\kappa$ which pass through x define two points $x^- < x^+$. Necessarily $[x^-, x^+] \subset]x_{n/m}, x'_{n/m}[$ for some $n/m \in \mathbb{Q}_\psi$ (because the existence of $y \in [x^-, x^+] \cap C_\psi$ would lead to the contradiction $x \notin \Delta_y^*$), hence $x \in \Delta_{n/m}$. Thus x belongs to the union of diamonds $\Delta_{n/m}$ in both cases and this yields the reverse inclusion.

As a consequence $C_{\psi,\kappa}$ is closed and its intersection with a strip $\left\{|\Im\mathrm{m}\, x| \le d\right\}$ is compact (see Figure 2). The inequalities

$$\forall n/m \in \mathbb{Q}_\psi, \quad \operatorname{meas}\left(\Delta_{n/m}\right) = \frac{1}{2}\kappa(x'_{n/m} - x_{n/m})^2 < 8\kappa\left(\tfrac{\psi(m)}{m}\right)^2 \le \frac{2\kappa}{m^4}$$

(which follow from Proposition 2.2 (2) and from $\psi(m) \le 1/2m$) yield the last statement. □

REMARK 2.5. Using a suggestion by Herman ([He], Remark at p. 81), one can prove that $\mathcal{O}(C_{\psi,\kappa,d}, B) = \mathcal{R}(C_{\psi,\kappa,d}, B)$, a result to be compared with the general inclusion which was indicated in Section 2.1. Notice that the relation

$$\mathbb{R} \subset \bigcap_{y \in C_\psi} \overline{\Delta_y^*} = C_\psi \cup \bigsqcup_{n/m \in \mathbb{Q}_\psi} \overline{\Delta_{n/m}}$$

implies that

$$\operatorname{int}(C_{\psi,\kappa,d}) = \{x \in \mathbb{C}/\mathbb{Z} \mid |\Im\mathrm{m}\, x| < d\} \setminus \Big(C_\psi \cup \bigsqcup_{n/m\in\mathbb{Q}_\psi} \overline{\Delta_{n/m}}\Big) \subset \mathbb{C}\setminus\mathbb{R}\ (\mathrm{mod}\,\mathbb{Z}).$$

The idea is to apply Milnikov's theorem [Za, p. 112] which states that, if the inner boundary of a compact set K is a subset of an analytic curve, $\mathcal{O}(K,B) = \mathcal{R}(K,B)$. (The inner boundary of K is defined as $\partial_I K = \partial K \setminus \bigsqcup \partial\Delta_\ell$, where $\bigsqcup \Delta_\ell$ is the decomposition of $\mathbb{C}\setminus K$ into disjoint connected components. Here $\partial_I C_{\psi,\kappa,d} = C_\psi \subset \mathbb{R}/\mathbb{Z}$.)

REMARK 2.6. One can check that $C_{\psi,\kappa,d}$ has a finite number of connected components and is locally connected; it is connected as soon as $d > \kappa\psi(2)$. Also $\operatorname{int}(C_{\psi,\kappa,d})$ has a finite number of connected components.

2.3.4. Finally we define the sequence of compact subsets K_j of $\mathbb{C}$ which will be used in the sequel.

DEFINITION 2.5. Let fix us a decreasing sequence $(\gamma_j)_{j\geq 0}$ wich tends to 0 and numbers $\kappa \in]0,1[$, $d, \alpha > 0$. We assume $\gamma_j < \inf(\frac{\alpha e}{2}, \frac{e^\alpha - 1}{2}, 1)$ for all $j \geq 0$. We define

$$\psi_j(m) = \gamma_j e^{-\alpha m} \text{ for } m \geq 1, \quad K_j = \{q = e^{2\pi i x},\ x \in C_{\psi_j,\kappa,d}\}, \quad C = \bigcup_{j\in\mathbb{N}} K_j.$$

Observe that each K_j is contained in the annulus $\{e^{-2\pi d} \leq |q| \leq e^{2\pi d}\}$ and that its measure tends to the measure of this annulus, while the measure of $K_j \cap \mathbb{S}^1$ tends to the measure of the circle, as $j \to \infty$.

REMARK 2.7. Since $C \cap \mathbb{S}^1 = \cup_{j\in\mathbb{N}}\{ e^{2\pi i x} \mid x \in C_{\psi_j}\}$, by Lemma 2.3 it is a countable union of nowhere dense closed sets. Proposition 2.2 then shows that its complement in $\mathbb{S}^1$ is a dense G_δ-set with zero s-dimensional Hausdorff measure for all $s > 0$.

LEMMA 2.5. *There exists a positive number μ, which depends only on κ, such that*

$$\forall j \in \mathbb{N},\ \forall q \in K_j,\ \forall \Lambda \in \mathcal{R}, \quad |q - \Lambda| > \mu\, \frac{\psi_j(m(\Lambda))}{m(\Lambda)}.$$

PROOF. Let $j \in \mathbb{N}$ and $q \in K_j$. Since $\frac{\psi_j(m)}{m} \leq \frac{1}{2}$ for all $m \geq 1$, we may suppose that $\operatorname{dist}(q, \mathbb{S}^1)$ be less than some arbitrary constant; thus we assume

$$q = e^{2\pi i x}, \quad x \in C_{\psi_j,\kappa}\ (\mathrm{mod}\,\mathbb{Z}), \quad |\Im\mathrm{m}\, x| \leq 1.$$

We also choose $y \in C_{\psi_j}\ (\mathrm{mod}\,\mathbb{Z})$ such that $|\Im\mathrm{m}\, x| \geq \kappa |\Re\mathrm{e}(x-y)|$.

Let $\Lambda \in \mathcal{R}$. We choose $n/m \in \mathbb{Q}$ such that $\Lambda = e^{2\pi i n/m}$ and $|\Re\mathrm{e}(x - \frac{n}{m})| \leq \frac{1}{2}$. According to (2.5), $|y - \frac{n}{m}| > \frac{\psi_j(m)}{2m}$, and one can check easily that $|x - \frac{n}{m}| \geq \mu_0 |y - \frac{n}{m}|$ with $\mu_0 = (1+\kappa^{-2})^{-1/2}$. Thus $z = x - \frac{n}{m}$ satisfies

$$|\Re\mathrm{e}\, z| \leq \frac{1}{2}, \quad |\Im\mathrm{m}\, z| \leq 1, \quad |z| > \mu_0 \frac{\psi_j(m)}{2m}.$$

Hence $|q - \Lambda| = |e^{2\pi i z} - 1|$ can be bounded from below as required. □

2.4. Monogenic regularity of the solutions

Let B a Banach space. We now consider the space of B-valued monogenic functions which corresponds to the sequence (K_j) of Definition 2.5. We will see that the general solution of the cohomological equation as encoded by the mapping F_{r_1,r_2} of Section 1.3 belongs to this space—recall its definition (1.4) and the notation $B_r = zH^\infty(\mathbb{D}_r)$; of course $B = \mathcal{L}(B_{r_1}, B_{r_2})$ in that case.

More generally, we will show that the Borel-Wolff-Denjoy series of the form

$$\Sigma_{\mathcal{R}}(a) : q \mapsto \big(\Sigma_{\mathcal{R}}(a)\big)(q) = \sum_{\Lambda\in\mathcal{R}} \frac{a_\Lambda}{q-\Lambda} \tag{2.10}$$

(not necessarily with the same coefficients as those of (2.3) in Remark 2.4) are monogenic; we simply restrict ourselves to

$$\mathcal{S}(r,B) = \Big\{ a = \{a_\Lambda\}_{\Lambda\in\mathcal{R}} \text{ sequence of } B \text{ such that } \exists c>0 \,/\, \forall\Lambda\in\mathcal{R}, \|a_\Lambda\| \le \tfrac{c\,r^{m(\Lambda)}}{m(\Lambda)} \Big\} \tag{2.11}$$

for some $r \in]0, e^{-3\alpha}[$.

THEOREM 2.2. *For all $r \in]0, e^{-3\alpha}[$ the Borel-Wolff-Denjoy series of the form $\Sigma_{\mathcal{R}}(a)$, $a \in \mathcal{S}(r,B)$, belong to $\mathcal{M}((K_j),B)$. In particular, this the case for the general solution F_{r_1,r_2} if $B = \mathcal{L}(B_{r_1}, B_{r_2})$ and $0 < r_2 < r_1\, e^{-3\alpha}$.*

PROOF. According to Definition 2.3 we must check that

$$f = \Sigma_{\mathcal{R}}(a) \in C^1_{hol}(K_j, B)$$

for every $a \in \mathcal{S}(r,B)$ and $j \in \mathbb{N}$. It is natural to define the function

$$f^{(1)}(q) = -\sum_{m=1}^{\infty} \sum_{\Lambda\in\mathcal{R}^*_m} \frac{a_\Lambda}{(q-\Lambda)^2}$$

whose restriction to $\mathrm{int}(K_j)$ is just the ordinary derivative of f.

According to Lemma 2.5,

$$\forall q \in K_j,\ \forall \Lambda \in \mathcal{R}, \quad |q-\Lambda| \ge \mu\gamma_j \frac{e^{-\alpha m(\Lambda)}}{m(\Lambda)}. \tag{2.12}$$

Thus, for $k = 0$ or 1, and for $q \in K_j$,

$$\|f^{(k)}(q)\| \le \sum_{m=1}^{\infty} \sum_{\Lambda\in\mathcal{R}^*_m} \frac{c\,r^m}{|q-\Lambda|^{k+1} m} \le c(\mu\gamma_j)^{-k-1} \sum_{m=1}^{\infty} m^{k+1} (r\,e^{(k+1)\alpha})^m < +\infty.$$

Note that f and $f^{(1)}$ are continuous since the convergence is uniform and K_j is compact. To prove C^1-smoothness, we consider the remainder

$$R(q,q') = \frac{f(q)-f(q')}{q-q'} - f^{(1)}(q') = -\sum_{m=1}^{\infty} \sum_{\Lambda\in\mathcal{R}^*_m} \frac{(q'-q)}{(q-\Lambda)(q'-\Lambda)^2} a_\Lambda.$$

Because of (2.12) and the assumption $r < e^{-3\alpha}$, we have $\|R(q,q')\| \le c_j |q-q'|$, with $c_j = c(\mu\gamma_j)^{-3} \sum_{m=1}^{\infty} m^3 (r\,e^{3\alpha})^m$. In particular Definition 2.1 is satisfied.

The statement about F_{r_1,r_2} is a particular case of what we just proved: choosing $a_\Lambda = \Lambda\mathcal{L}_{m(\Lambda)}\odot$ and $r = \frac{r_2}{r_1}$, we use Lemma A1.1 and see that $\|a_\Lambda\|_{\mathcal{L}(B_{r_1},B_{r_2})} \leq \|\mathcal{L}_{m(\Lambda)}\|_{B_r} \simeq \frac{1}{m(\Lambda)} r^{m(\Lambda)}$.

□

As for the fundamental solution, notice that $f_\delta \in \mathcal{M}((K_j), B_r)$ as soon as $0 < r < e^{-3\alpha}$.

2.5. Whitney smoothness of monogenic functions

As already mentioned in Remark 2.4, we cannot apply Borel's Theorem to conclude that functions in $\mathcal{M}((K_j)_{j\in\mathbb{N}}, B)$ are $\mathcal{C}^\infty$-holomorphic in some subsets of the K_j's. But this can be shown directly.

Let $c_0(\mathbb{R})$ denote the classical Banach space of real sequences $s = (s_k)_{k\geq 0}$ such that $s_k \to 0$ as $k \to +\infty$, endowed with the norm $\|s\| = \sup|s_k|$. A subset A of $c_0(\mathbb{R})$ is *closed* and *totally bounded* if and only if the following two conditions are satisfied:
(i) $\exists C > 0 \;/\; \forall s \in A,\ \|s\| \leq C.$
(ii) $\forall \varepsilon > 0,\ \exists k_0 \in \mathbb{N} \;/\; \forall s \in A,\ \forall k \geq k_0,\ |s_k| \leq \varepsilon.$

DEFINITION 2.6. To $\gamma \in]0,1[$ and A, totally bounded closed subset of $c_0(\mathbb{R})$, we associate

$$W_\gamma^A = \{x \in \mathbb{R}\setminus\mathbb{Q}\,(\mathrm{mod}\,\mathbb{Z}) \mid \exists s \in A \text{ such that } \forall k \in \mathbb{N},\ m_{k+1}(x) \leq \gamma^{-1} e^{s_k m_k(x)}\}.$$

If moreover $\kappa \in]0,1[$ and $d > 0$, we define

$$W_{\gamma,\kappa,d}^A = \bigcup_{y \in W_\gamma^A} \{x \in \mathbb{C}/\mathbb{Z} \mid \kappa|\Re\mathrm{e}(\tilde{x}-\tilde{y})| \leq |\Im\mathrm{m}\,\tilde{x}| \leq d\},$$

where $\tilde{x}$ and $\tilde{y}$ denote some lifts in $\mathbb{C}$ of x and y.

One can study the sets W_γ^A and $W_{\gamma,\kappa,d}^A$ with the same kind of arguments as in Section 2.3. For instance one can easily check that they are closed and perfect. Notice that W_γ^A is non-empty as soon as A contains a sequence s such that $s_k \geq 2G^{\frac{3-k}{2}}$ for all k (indeed $g \in W_\gamma^A$ in that case). Moreover, if $x \in \mathbb{R}\setminus\mathbb{Q}\,(\mathrm{mod}\,\mathbb{Z})$ satisfies the condition

$$\lim_{k\to\infty} \frac{\log m_{k+1}(x)}{m_k(x)} = 0, \tag{2.13}$$

and if $\alpha > 0$ is given, there exist $\gamma \in]0,1[$ and $s \in c_0(\mathbb{R})$ such that $x \in W_\gamma^{\{s\}}$ and $\|s\| \leq \alpha$.

THEOREM 2.3. *Let $\gamma, \kappa \in]0,1[$, $d > 0$, ψ an approximation function of the form $\psi(m) = \gamma e^{-\alpha m}$ and $K = \{q = e^{2\pi i x},\ x \in C_{\psi,\kappa,d}\}$. Let A a totally bounded closed subset of $c_0(\mathbb{R})$ such that $\forall s \in A$, $\|s\| \leq \alpha$, and $K^* = \{q = e^{2\pi i x},\ x \in W_{8\gamma,\kappa,d/2}^A\}$. Then $K^* \subset K$ and $\mathcal{C}_{hol}^1(K, B) \subset \mathcal{C}_{hol}^\infty(K^*, B)$ for any Banach space B.*

PROOF. It is immediate to check that $W_{8\gamma}^A \subset C_\psi = \{x \mid \forall k \in \mathbb{N},\ m_{k+1}(x) \leq \gamma^{-1}\, e^{\alpha m_k(x)}\}$; thus $K^* \subset K$.

Let $f \in C^1_{hol}(K, B)$. We will use Remark 2.1. Observe that, in view of Lemma 2.4, the connected components of $(\mathbb{C}/\mathbb{Z}) \setminus C_{\psi,\kappa,d}$ are of the form $\Delta_{n/m}$ with $n/m \in \mathbb{Q}_\psi$, except for one or two of them: the components of $i\infty$ and $-i\infty$ may be reduced to the half-planes $\{\pm \Im m\, x > d\}$, or else they both coincide with the union of these half-planes and a finite number of diamonds $\Delta_{n/m}$. From that we deduce the decomposition $\bigsqcup_{\ell \geq 1} U_\ell$ of $\mathbb{C} \setminus K$ into connected components—the index $\ell = 1$ (resp. $\ell = 1$ and 2) will correspond to the exceptional component (resp. components), the next ones being numbered as $U_\ell = \exp(2\pi i \Delta_{n_\ell/m_\ell})$ with a non-decreasing sequence (m_ℓ).

Moreover, for each $n/m \in \mathbb{Q}_\psi$, we recall that according to Proposition 2.2,

$$\left| X - \frac{n}{m} \right| < r_{n/m} = \frac{2\gamma}{m}\, e^{-\alpha m} \quad \text{if } X = x_{n/m} \text{ or } x'_{n/m},$$

hence $\partial\Delta_{n/m}$ has length less than $4 r_{n/m}\sqrt{1+\kappa^2}$. The series $\sum \operatorname{length}(\partial U_\ell)$ is thus convergent.

Let $j \in \mathbb{N}$. We will now check that the series

$$\sum_{\ell \geq 1} \int_{\partial U_\ell} \frac{|d\zeta|}{|\zeta - q|^{j+1}} \tag{2.14}$$

is uniformly convergent for $q \in K^*$. This will allow us to set

$$f^{(j)}(q) = \frac{j!}{2\pi i} \sum_{\ell \geq 1} \int_{\partial U_\ell} \frac{f(\zeta)}{(\zeta - q)^{j+1}}\, d\zeta. \tag{2.15}$$

LEMMA 2.7. *There exists a positive number μ (which depends only on κ) such that, whenever $n/m \in \mathbb{Q}_\psi$,*

$$\forall \xi \in \overline{\Delta}_{n/m}, \quad \operatorname{dist}(e^{2\pi i\xi}, K^*) > \frac{2\mu\gamma}{m}\, e^{-\alpha m}.$$

For each $j \in \mathbb{N}$, there exists a positive integer M (which depends only on γ, α and j) such that, whenever $n/m \in \mathbb{Q}_\psi$ and $m \geq M$,

$$\forall \xi \in \overline{\Delta}_{n/m}, \quad \operatorname{dist}(e^{2\pi i\xi}, K^*) > \frac{2\mu\gamma}{m}\, e^{-\frac{\alpha m}{2(j+1)}}. \tag{2.16}$$

We end the proof of Theorem 2.3 before proving Lemma 2.7. According to the first part of Lemma 2.7, each term in the series (2.14) is well defined when $q \in K^*$. For ℓ large enough (say $\ell \geq L$), $U_\ell = \exp(2\pi i \Delta_{n_\ell/m_\ell})$ with $n_\ell/m_\ell \in \mathbb{Q}_\psi$ and $m_\ell \geq M$, thus we can use (2.16) for each $q \in K^*$:

$$\begin{aligned}\int_{\partial U_\ell} \frac{|d\zeta|}{|\zeta - q|^{j+1}} &\leq 2\pi\, e^{2\pi d} \Big(\frac{m_\ell}{2\mu\gamma}\Big)^{j+1} e^{\frac{\alpha m_\ell}{2}} \operatorname{length}(\partial\Delta_{n_\ell/m_\ell}) \\ &\leq \frac{8\pi\, e^{2\pi d}\sqrt{1+\kappa^2}}{\mu} \Big(\frac{m_\ell}{2\mu\gamma}\Big)^{j} e^{-\frac{\alpha m_\ell}{2}}.\end{aligned}$$

The series (2.14) is thus convergent, and we can use (2.15) with $j = 0$ or 1 to represent f or $f^{(1)}$ in K^*. For $j \geq 2$, we define $f^{(j)}$ in K^* by (2.15), and the previous computation shows the existence of $C > 0$ such that

$$\forall \ell \geq L,\ \forall q \in K^*, \quad \Big\| \int_{\partial U_\ell} \frac{f(\zeta)}{(\zeta - q)^{j+1}}\, d\zeta \Big\| \leq C\, m_\ell^j\, e^{-\frac{\alpha m_\ell}{2}}$$

(and for $\ell < L$ this expression is continuous in q); hence, by uniform convergence, $f^{(j)}$ is continuous in K^*.

Let us consider the Taylor remainders

$$R^{(j,v)}(q,q') = f^{(j)}(q') - \sum_{u=0}^{v} \frac{1}{u!} f^{(j+u)}(q)(q'-q)^u$$

for $j, v \geq 0$ and $q, q' \in K^*$. Remark 2.5 applies also to $W^A_{8\gamma,\kappa,d/2}$, and thus to K^*: these sets have a finite number of connected components and are locally connected. In fact, for $q, q' \in K^*$ close enough (say $|q - q'| \leq \delta$), one can define a path $\Gamma(q,q')$ which joins q to q' inside K^* and which is the image by $x \mapsto e^{2\pi i x}$ of the union of 1, 2 or 3 segments of slopes $\pm\kappa$; the length of $\Gamma(q,q')$ is less than $\nu|q'-q|$, where ν depends only on κ.

We now conclude the proof of Theorem 2.3 by checking that there exists $C > 0$ such that

$$\forall q, q' \in K^*, \quad |q - q'| \leq \delta \;\Rightarrow\; \|R^{(j,v)}(q,q')\| \leq C\,|q'-q|^{v+1}. \tag{2.17}$$

We can write

$$R^{(j,v)}(q,q') = \frac{j!}{2\pi i} \sum_{\ell \geq 1} \int_{\partial U_\ell} \mathcal{R}^{(j,v)}(q,q',\zeta) f(\zeta)\, d\zeta,$$

where $\mathcal{R}^{(j,v)}(q,q',\zeta)$ is the Taylor remainder at order v for the function $q' \mapsto (\zeta - q')^{-j-1}$, i.e.

$$\mathcal{R}^{(j,v)}(q,q',\zeta) = \frac{(j+v+1)!}{j!\,v!} \int_{\Gamma(q,q')} \frac{(q'-q'')^v}{(\zeta - q'')^{j+v+2}}\, dq''.$$

From this identity and from Lemma 2.7 applied with j replaced by $j+v+1$, one can deduce the existence of a positive integer L such that, if $\zeta \in \partial U_\ell$ with $\ell \geq L$,

$$\|\mathcal{R}^{(j,v)}(q,q',\zeta)\| \leq \text{const}\; m_\ell^{j+v+2}\, e^{\frac{\alpha m_\ell}{2}}\, |q-q'|^{v+1},$$

whereas for $\ell < L$, $\|\mathcal{R}^{(j,v)}(q,q',\zeta)\| \leq \text{const}\, |q-q'|^{v+1}$. Therefore, the validity of (2.17) follows from the inequalities $\text{length}(\partial U_\ell) \leq \frac{\text{const}}{m_\ell}\, e^{-\alpha m_\ell}$. □

Proof of Lemma 2.7. We must show that $|q - e^{2\pi i\xi}| > \text{const}\frac{\psi(m)}{m}$ for $q \in K^*$ and $\xi \in \overline{\Delta}_{n/m}$. Notice that $\frac{\psi(m)}{m} \leq \gamma\, e^{-\alpha} \leq \frac{1}{2}$ since ψ is an approximation function, and $|\Im\text{m}\,\xi| \leq \kappa(x'_{n/m} - x_{n/m}) < \kappa$. Therefore we can assume

$$q = e^{2\pi i x} \quad \text{with } x \in W^A_{8\kappa,\kappa,d}, \quad |\Im\text{m}\, x| \leq 2\kappa.$$

Moreover we can consider that $|\Re\text{e}\, z| \leq \frac{1}{2}$, where $z = x - \xi$, and since $|q - e^{2\pi i\xi}| \geq e^{-4\pi\kappa}\,|1 - e^{-2\pi i z}|$, it will be enough to bound from below $|z|$ itself (z lies indeed in a domain where $|(e^{-2\pi i z} - 1)/z|$ is bounded from below). The same reasoning holds for the proof of (2.16) provided that we take $m \geq M \geq 2(j+1)$.

In fact we will prove the inequalities

$$\forall n/m \in \mathbb{Q}_\psi,\ \forall y \in W^A_{8\gamma}, \quad \left| y - \frac{n}{m} \right| \geq \frac{4\gamma}{m}\, e^{-\alpha m}, \tag{2.18}$$

and the existence, for each $j \in \mathbb{N}$, of a positive integer M such that

$$(2.19) \qquad \forall n/m \in \mathbb{Q}_\psi,\ \forall y \in W^A_{8\gamma}, \quad m \geq M \ \Rightarrow\ \left| y - \frac{n}{m} \right| \geq \frac{4\gamma}{m}\, e^{-\frac{\alpha m}{2(j+1)}}.$$

This is enough to bound $|z| = |x - \xi|$ from below as required since for any $x \in W^A_{8\gamma,\kappa,d}$ there exists $y \in W^A_{8\gamma}$ such that $|\Im\mathrm{m}\, x| \geq \kappa |\Re\mathrm{e}(x-y)|$, but then $|x - \xi| \geq (1+\kappa^2)^{-1/2} \operatorname{dist}\big(y, [x_{n/m}, x'_{n/m}]\big)$ for all $\xi \in \overline{\Delta}_{n/m}$, and

$$\operatorname{dist}\big(y, [x_{n/m}, x'_{n/m}]\big) \geq |y - \tfrac{n}{m}| - \max\big(x'_{n/m} - \tfrac{n}{m}, |x_{n/m} - \tfrac{n}{m}|\big) < |y - \tfrac{n}{m}| - \tfrac{2\gamma}{m}\, e^{-\alpha m}.$$

Let $y \in W^A_{8\gamma}$. Let $s \in A$ such that $m_{k+1}(y) \leq \frac{1}{8\gamma}\, e^{s_k m_k(y)}$. According to (A3.4),

$$\forall k \geq 0, \qquad \left| y - \frac{n_k(y)}{m_k(y)} \right| > \frac{4\gamma}{m_k(y)}\, e^{-s_k m_k(y)} \geq \frac{4\gamma}{m_k(y)}\, e^{-\alpha m_k(y)}.$$

Let $n/m \in \mathbb{Q}_\psi$. Either $\frac{n}{m} = \frac{n_k(y)}{m_k(y)}$ for some $k \geq 0$ and (2.18) is proved. Or $\frac{n}{m}$ is not a convergent of y; then $m_{k-1}(y) \leq m < m_k(y)$ for some $k \geq 1$ and Proposition A3.4 applies:

$$m \left| y - \frac{n}{m} \right| > m_{k-1}(y) \left| y - \frac{n_{k-1}(y)}{m_{k-1}(y)} \right| \geq 4\gamma\, e^{-s_{k-1} m_{k-1}(y)} \geq 4\gamma\, e^{-\alpha m}.$$

Therefore (2.18) is true in all cases.

As for (2.19), given $j \in \mathbb{N}$ we first choose $k_0 \geq 0$ such that $|s_k| \leq \frac{\alpha}{2(j+1)}$ for all $k \geq k_0$ and $s \in A$. We then choose $M \geq 1$ such that

$$\forall y \in W^A_{8\gamma}, \quad m_{k_0+1}(y) < M.$$

(For all $y \in W^A_{8\gamma}$, $m_0(y) = 1$ thus $m_1(y) \leq \frac{1}{8\gamma}\, e^{\alpha} = M_1$, $m_2(y) \leq \frac{1}{8\gamma}\, e^{\alpha M_1} = M_2, \ldots$: take $M > M_{k_0+1}$.) According to (A3.4), we have now

$$\forall y \in W^A_{8\gamma},\ \forall k \geq k_0, \qquad \left| y - \frac{n_k(y)}{m_k(y)} \right| > \frac{4\gamma}{m_k(y)}\, e^{-\frac{\alpha m_k(y)}{2(j+1)}}.$$

Let $y \in W^A_{8\gamma}$ and $n/m \in \mathbb{Q}_\psi$ with $m \geq M$. We are faced with the same alternative as above, but we know moreover that if $\frac{n}{m} = \frac{n_k(y)}{m_k(y)}$ or $m_{k-1}(y) \leq m < m_k(y)$, necessarily $k \geq k_0 + 1$. Therefore we obtain the refined inequality (2.19) in all cases. □

DEFINITION 2.7. For any closed totally bounded subset A of $c_0(\mathbb{R})$ and any integer j, we define

$$K^*_{A,j} = \{q = e^{2\pi i x},\ x \in W^A_{8\gamma_j,\kappa,d/2}\}$$

provided that $\|s\| \leq \alpha$ for all $s \in A$, with the same notations as in Definition 2.5.

According to Theorem 2.3, $K^*_{A,j} \subset K_j$ and $\mathcal{C}^1_{hol}(K_j, B) \subset \mathcal{C}^\infty_{hol}(K^*_{A,j}, B)$. In particular, according to Theorem 2.2, the solutions of the cohomological equation are $\mathcal{C}^\infty$-holomorphic in each $K^*_{A,j}$.

Observe that any point of the form $\lambda = e^{2\pi i x}$ with x satisfying (2.13) lies in $K^*_{\{s\},j}$ for s well chosen and j large enough.

CHAPTER 3

Carleman Classes at Diophantine Points

In this section, we address the following question (directly inspired by [He], Question at p. 82): Do the solutions of the cohomological equation belong to any quasianalytic Carleman class? We will treat separately some particular points of $\mathbb{S}^1$ among those at which Theorem 2.3 yields Whitney smoothness, and study asymptotic expansions in disks tangent to $\mathbb{S}^1$ at each of these points.

As a preliminary, in Section 3.1, we define the Carleman classes $\mathcal{C}^{\pm}(\lambda, \{M_n\}, B)$ which we think are the most relevant for the problem at hand[3]. We recall a well-known criterium of quasianalyticity due to Carleman, and we also introduce spaces of functions which admit Gevrey asymptotic expansions. Our presentation is somewhat influenced by the works of Ramis and Malgrange on divergent series (see for instance [Ra], [Ma]).

In Section 3.2 we prove that all functions monogenic in the compacts K_j of Definition 2.5 admit Gevrey-τ asymptotic expansions at Diophantine points of exponent $\tau \geq 2$. On the other hand, in the case of the fundamental solution, we prove in Section 3.3 the sharpness of the index $\tau = 2$ in Gevrey asymptotics for those Diophantine points which correspond to quadratic irrationals, and conclude that no quasianalytic Carleman class at those points contains the fundamental solution.

3.1. Carleman and Gevrey classes

3.1.1. Let B be a complex Banach space, whose norm we denote by $\| \,.\, \|$, and $\lambda \in \mathbb{S}^1$. Let us fix some sequence $\{M_n\}_{n\geq 0}$ of positive numbers.

DEFINITION 3.1. We define the Carleman class $\mathcal{C}^-(\lambda, \{M_n\}, B)$ to be the vector space of all B-valued functions f for which there exist an open disk $\Delta \subset \mathbb{D}$ tangent to $\mathbb{S}^1$ at λ (see Figure 3), a formal series $\sum_{n\geq 0} a_n Q^n \in B[[Q]]$ and positive numbers c_0 and c_1 such that the function f is holomorphic in Δ and

$$\forall N \geq 0,\ \forall q \in \Delta, \quad \|f(q) - \sum_{0\leq n\leq N-1} a_n (q-\lambda)^n\| \leq c_0\, c_1^N\, M_N\, |q-\lambda|^N.$$

The mapping

$$J_\lambda^- : f \in \mathcal{C}^-(\lambda, \{M_n\}, B) \ \mapsto \ \sum_{n\geq 0} a_n Q^n \in B[[Q]]$$

[3] Carleman classes are usually defined as spaces of functions which are defined and C^∞ (in the real sense) on some—possibly infinite—interval I of $\mathbb{R}$ and whose derivatives satisfy some uniform bounds (see [Th]); the relationship between such classes with $I = \mathbb{R}^+$ and the spaces $\mathcal{C}^{\pm}(\lambda, \{M_n\}, B)$ defined below is indicated in [Ca].

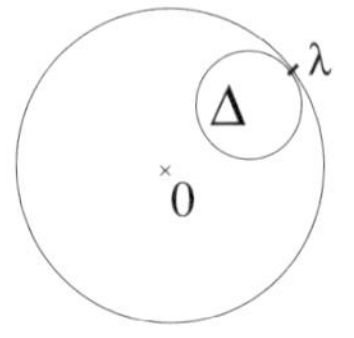

FIGURE 3

which associates to a function of $\mathcal{C}^-(\lambda, \{M_n\}, B)$ its asymptotic expansion at λ is well defined. In fact the functions of $\mathcal{C}^-(\lambda, \{M_n\}, B)$ are $\mathcal{C}^\infty$-holomorphic in the sense of Definition 2.2:

LEMMA 3.1. *If $f \in \mathcal{C}^-(\lambda, \{M_n\}, B)$ and $S \subset \mathbb{D}$ is a bounded sector of vertex λ and small enough radius, there exist positive numbers c_0 and c_1 such that*
- *the function f is $\mathcal{C}^\infty$-holomorphic in the closure $\bar{S}$ of the sector,*
- *for all $n \geq 0$ and $q \in \bar{S}$, $\|\frac{1}{n!} f^{(n)}(q)\| \leq c_0\, c_1^n\, M_n$,*
- *and $J_\lambda^-(f) = \sum_{n\geq 0} \frac{1}{n!} f^{(n)}(\lambda)\, Q^n$.*

Conversely, if a function f is $\mathcal{C}^\infty$-holomorphic in a closed disk $\bar{\Delta} \subset \overline{\mathbb{D}}$ tangent to $\mathbb{S}^1$ at λ and satisfies inequalities of the form $\|\frac{1}{n!} f^{(n)}(q)\| \leq c_0\, c_1^n\, M_n$ in $\bar{\Delta}$, then it belongs to $\mathcal{C}^-(\lambda, \{M_n\}, B)$.

Here, by "bounded sector" we mean the intersection of an open infinite sector and of some open disk centered at its vertex, and we call "radius" of the sector the radius of that disk.

We leave the proof of Lemma 3.1 to the reader (one can use the Taylor-Lagrange formula).

As a consequence, the asymptotic expansion $J_\lambda^-(f)$ of any $f \in \mathcal{C}^-(\lambda, \{M_n\}, B)$ belongs to the space $B[[Q]]_{\{M_n\}}$ defined as

$$\Big\{ \sum_{n\geq 0} a_n Q^n \in B[[Q]] \mid \exists c_0, c_1 > 0 \text{ such that } (\forall n \geq 0)\ \|a_n\| \leq c_0\, c_1^n\, M_n \Big\}.$$

By definition, the space $\mathcal{C}^-(\lambda, \{M_n\}, B)$ is quasianalytic at λ if and only if the mapping J_λ^- is injective.

Analogously we define the space $\mathcal{C}^+(\lambda, \{M_n\}, B)$ by using disks Δ contained in $\mathbb{E}$ instead of $\mathbb{D}$, and the corresponding mapping

$$J_\lambda^+ : \mathcal{C}^+(\lambda, \{M_n\}, B) \to B[[Q]]_{\{M_n\}}.$$

The change of variable $q \mapsto \lambda^2/q$ induces an isomorphism between $\mathcal{C}^-(\lambda, \{M_n\}, B)$ and $\mathcal{C}^+(\lambda, \{M_n\}, B)$.

We can now state Carleman's criterium of quasianalyticity [Ca]:

THEOREM 3.1 (Carleman's Criterium). *The space $\mathcal{C}^\pm(\lambda, \{M_n\}, B)$ is quasianalytic at λ (i.e. $J_\lambda^\pm$ is injective on that space) if and only if $\sum_{n\geq 1} \frac{1}{\beta_n} = +\infty$ where*

$$\beta_n = \inf_{n'\geq n} \{ M_{n'}^{1/n'} \}.$$

REMARK 3.1. The criterium is usually stated for spaces of scalar functions, but it is also valid for spaces of B-valued functions (as soon as $B \neq \{0\}$ of course). The quasianalyticity of $\mathcal{C}^{\pm}(\lambda, \{M_n\}, B)$ is indeed equivalent to that of $\mathcal{C}^{\pm}(\lambda, \{M_n\}, \mathbb{C})$ because of the existence of non-trivial continuous linear functionals on any normed linear space: if f is a function in $\mathcal{C}^{\pm}(\lambda, \{M_n\}, B)$, any continuous linear functional ℓ on B induces a function $\ell \circ f$ which belongs to $\mathcal{C}^{\pm}(\lambda, \{M_n\}, \mathbb{C})$, and $J_\lambda^{\pm}(\ell \circ f) = \ell\big(J_\lambda^{\pm}(f)\big)$ (letting ℓ act termwise in $B[[Q]]$ in order to define the right-hand side). The point is that for a function f to be identically zero, it is necessary and sufficient that all the functions $\ell \circ f$ vanish identically (given any Banach space B its dual separates points on B).

Let $\mathcal{C}(\lambda, \{M_n\}, B)$ be the space of all B-valued functions for which there exist disks $\Delta^- \subset \mathbb{D}$ and $\Delta^+ \subset \mathbb{E}$ tangent to $\mathbb{S}^1$ at λ such that $f_{|\Delta^-} \in \mathcal{C}^-(\lambda, \{M_n\}, B)$, $f_{|\Delta^+} \in \mathcal{C}^+(\lambda, \{M_n\}, B)$ and $J_\lambda^-(f_{|\Delta^-}) = J_\lambda^+(f_{|\Delta^+})$. We will denote by $J_\lambda(f)$ simply the asymptotic expansion at λ of a function f of $\mathcal{C}(\lambda, \{M_n\}, B)$. As a consequence of Theorem 3.1, $\mathcal{C}(\lambda, \{M_n\}, B)$ is quasianalytic at any point of $\mathbb{D} \cup \{\lambda\} \cup \mathbb{E}$ if and only if $\sum \frac{1}{\beta_n} = +\infty$.

3.1.2. As a special case we will consider *Gevrey classes*, i.e. spaces of functions with Gevrey-τ asymptotic expansion for some $\tau \geq 0$.

DEFINITION 3.2. If B is a Banach space, $\lambda \in \mathbb{S}^1$ and $\tau \in \mathbb{R}^+$, we define the Gevrey classes

$$\mathcal{G}_\tau^-(\lambda, B), \qquad \mathcal{G}_\tau^+(\lambda, B), \qquad \mathcal{G}_\tau(\lambda, B)$$

respectively as the Carleman classes

$$\mathcal{C}^-(\lambda, \{M_n\}, B), \;\; \mathcal{C}^+(\lambda, \{M_n\}, B), \;\; \mathcal{C}(\lambda, \{M_n\}, B)$$

with the sequence $\{M_n = \Gamma(1 + n\tau)\}$. We also set $B[[Q]]_\tau = B[[Q]]_{\{M_n\}}$ with the same sequence $\{M_n\}$.

We warn the reader that not all the authors follow this convention for indexing Gevrey classes. For us, $\tau = 0$ corresponds to the analytic class: $B[[Q]]_0$ is the space $B\{Q\}$ of convergent series, and J_λ^- and J_λ^+ are isomorphisms in that case. Thus $\mathcal{G}_0^{\pm}(\lambda, B)$ and $\mathcal{G}_0(\lambda, B)$ can all be identified to the space of all germs of B-valued holomorphic functions at λ.

We retain that, by Carleman's Theorem, the space $\mathcal{G}_\tau^{\pm}(\lambda, B)$ is quasianalytic at λ if and only if $\tau \leq 1$; and the same is true for $\mathcal{G}_\tau(\lambda, B)$. One can check that, if B is a Banach algebra, $\mathcal{G}_\tau^{\pm}(\lambda, B)$ and $\mathcal{G}_\tau(\lambda, B)$ are in fact algebras: they are stable by multiplication [Ma].

3.1.3. We will now focus on the $\tau = 1$ case and the relationship with the Laplace transform. We suppose moreover that B is a Banach algebra.

We denote by $\hat{\mathcal{N}}^{\pm}(B)$ the space of all B-valued functions $\hat{\phi}$ for which there exist some positive numbers $\rho' < \rho$ and some real number δ such that $\hat{\phi}$ is holomorphic in the open "half-strip"

$$H_\rho^{\pm} = \{\, \xi \in \mathbb{C} \mid \operatorname{dist}(\xi, \mathbb{R}^{\pm}) < \rho \,\}$$

and $\xi \mapsto e^{-\delta|\xi|}\, \|\hat{\phi}(\xi)\|$ is bounded in the closed half-strip $\bar{H}_{\rho'}^{\pm}$. The space $\hat{\mathcal{N}}^{\pm}(B)$ is stable by convolution, the convolution of two holomorphic functions $\hat{\phi}_1$ and $\hat{\phi}_2$ being defined as $\hat{\phi}_1 * \hat{\phi}_2(\xi) = \int_0^\xi \hat{\phi}_1(\xi_1)\hat{\phi}_2(\xi - \xi_1)\, d\xi_1$.

We also introduce a symbol δ_0 which one may think of as the Dirac distribution at the origin: identifying any pair $(a_0, \hat{\phi}) \in B \times \hat{\mathcal{N}}^{\pm}(B)$ with the symbolic sum $a_0\delta_0 + \hat{\phi} \in B\delta_0 \oplus \hat{\mathcal{N}}^{\pm}(B)$ and extending the convolution to the space $B\delta_0 \oplus \hat{\mathcal{N}}^{\pm}(B)$ by treating δ_0 as a unit, we get an algebra. The following theorem is due to Nevanlinna [Ma]:

THEOREM 3.2 (Nevanlinna's Theorem). *The Laplace transform*

$$\mathbb{L}_{\lambda}^{\pm} \,:\, a_0\delta_0 + \hat{\phi} \;\mapsto\; f^{\pm} \textit{ such that } f^{\pm}(\lambda(1+t)) = a_0 + \int_0^{\pm\infty} \hat{\phi}(\xi)\, e^{-\xi/t}\, d\xi$$

defines an isomorphism between the algebras $B\delta_0 \oplus \hat{\mathcal{N}}^{\pm}(B)$ *and* $\mathcal{G}_1^{\pm}(\lambda, B)$.

REMARK 3.2. Again we mention that the replacement of scalar functions by B-valued functions, with respect to the usual statement, is innocuous. Notice that Nevanlinna's Theorem implies that $\mathcal{G}_1^{\pm}(\lambda, B)$ is a differentiable algebra: it is stable by derivation (see Appendix A.5 for a description of the counterpart in the convolutive model $\hat{\mathcal{N}}^{\pm}(B)$ of such elementary operations as differentiation). Also, with respect to the notations of Definition 3.1, we have incorporated in our statement the change of infinitesimal variable $Q = q - \lambda \mapsto t = \lambda^{-1}Q$ in order to deal with Laplace integrals on $\mathbb{R}^{\pm}$ only (the counterpart in $\hat{\mathcal{N}}^{\pm}$ of such homotheties and of more general changes of variable is described in Appendix A.5).

Theorem 3.2 shows that the quasianalyticity of $\mathcal{G}_1^{\pm}(\lambda, B)$ is in some sense constructive, the reciprocal operator of $J_{\lambda}^{\pm}$ being described in terms of *Borel-Laplace resummation*:

DEFINITION 3.3. If $\tilde{f} = \sum a_n Q^n \in B[[Q]]_1$, we define a formal series $\tilde{\phi}(t) = \sum \phi_n t^n \in B[[t]]_1$ by $\tilde{\phi}(t) = \tilde{f}(\lambda t)$, and its *formal Borel transform* by $\phi_0\delta_0 + \hat{\phi}$ where

$$\hat{\phi}(\xi) = \sum_{n\geq 0} \phi_{n+1}\frac{\xi^n}{n!} = \sum_{n\geq 0} \lambda^{n+1} a_{n+1}\frac{\xi^n}{n!} \in B\{\xi\};$$

the series $\tilde{f}$ belongs to $J_{\lambda}^{\pm}\big(\mathcal{G}_1^{\pm}(\lambda, B)\big)$ if and only if $\hat{\phi}$ can be analytically continued to an element of $\hat{\mathcal{N}}^{\pm}(B)$, and its preimage is then equal to $\mathbb{L}_{\lambda}^{\pm}(a_0\delta_0 + \hat{\phi})$: it is called the *Borel-Laplace sum* of $\tilde{f}$ (in the direction of $\mathbb{R}^{\pm}$).

The reader is referred to Appendix A.5 for more details on the Borel-Laplace summation process.

DEFINITION 3.4. Let $\hat{\mathcal{N}}(B) = \hat{\mathcal{N}}^{-}(B) \cap \hat{\mathcal{N}}^{+}(B)$. We define $\mathbb{L}_{\lambda}$ in $B\delta_0 \oplus \hat{\mathcal{N}}(B)$ by gluing $\mathbb{L}_{\lambda}^{-}$ and $\mathbb{L}_{\lambda}^{+}$: we obtain an isomorphism between $B\delta_0 \oplus \hat{\mathcal{N}}(B)$ and $\mathcal{G}_1(\lambda, B)$.

3.2. Gevrey asymptotics at Diophantine points for monogenic functions

Let B a Banach space. According to Theorem 2.3, the monogenic functions belonging to $\mathcal{M}((K_j), B)$ are $\mathcal{C}^{\infty}$-holomorphic in the compacts $K_{A,j}^{*}$, with the notations of Definitions 2.5 and 2.7. In particular, any such function admits as asymptotic expansion its Taylor series at any point of $K_{A,j}^{*} \cap \mathbb{S}^1$. Among those points, some of them have further arithmetic properties which will yield Gevrey asymptotic expansions.

DEFINITION 3.5. Let $\gamma > 0$, $\tau \geq 2$. We define $\mathrm{DC}(\gamma,\tau)$ to be the set of all irrational numbers y which satisfy Diophantine inequalities of constant γ and exponent τ, i.e.

$$\forall n/m \in \mathbb{Q}, \quad |y - n/m| \geq \gamma\, m^{-\tau}.$$

We also set $\mathrm{DC}_\tau = \bigcup_{\gamma>0} \mathrm{DC}(\gamma,\tau)$ and $\underline{\mathrm{DC}}_\tau = \{\lambda = e^{2\pi i y},\ y \in \mathrm{DC}_\tau\}$.

It is well-known that DC_τ has full measure as soon as $\tau > 2$ and that DC_2 (which has measure zero) coincides with the set of constant-type irrationals (irrationals with bounded quotients—see the appendix, Paragraph A.3.3). For any $y \in \mathrm{DC}(\gamma,\tau)$, the property

$$\forall k \geq 0, \quad m_{k+1}(y) < \gamma^{-1}\, m_k(y)^{\tau-1}$$

(obvious consequence of Remark A3.2 in the appendix) allows one to find A and γ' such that $y \in W^A_{\gamma'}$ (e.g. $A = \{s\}$ with $s_k = m_k(y)^{-\delta}$ for some $\delta \in]0,1[$, and $\gamma' = \gamma \min_{k\geq 0}\{m_k(y)^{1-\tau}\exp(m_k(y)^{1-\delta})\}$). In particular, each point of $\underline{\mathrm{DC}}_\tau$ is contained in some $K^*_{A,j}$.

THEOREM 3.3. *Let $\tau \geq 2$. If $\lambda \in \underline{\mathrm{DC}}_\tau$, monogenic functions of $\mathcal{M}((K_j),B)$ admit Gevrey-τ asymptotic expansions at λ:*

$$\mathcal{M}((K_j),B) \subset \mathcal{G}_\tau(\lambda,B).$$

In particular, according to Theorem 2.2, the solution F_{r_1,r_2} of the cohomological equation belongs to $\mathcal{G}_\tau(\lambda,\mathcal{L}(B_{r_1},B_{r_2}))$ as soon as $0 < r_2 < r_1$ (using the fact that the positive number α which enters into the definition of the sequence (K_j) can be chosen arbitrarily small). Similarly $f_\delta \in \mathcal{G}_\tau(\lambda,B_r)$ if $0 < r < 1$.

The proof of Theorem 3.3 is somewhat analogous to that of Theorem 2.3. We first state a lemma about the relation between Diophantine points and the geometry of the compacts K_j, which parallels Lemma 2.7.

LEMMA 3.2. *Let $\tau \geq 2$ and $\lambda \in \underline{\mathrm{DC}}_\tau$. There exist $\mu > 0$, $j \geq 1$ and two open disks $\Delta^- \subset \mathbb{D}$ and $\Delta^+ \subset \mathbb{E}$ tangent to $\mathbb{S}^1$ at λ such that the set $\Delta^- \cup \{\lambda\} \cup \Delta^+$ is contained in K_j and, for every $n/m \in \mathbb{Q}_{\psi_j}$ and $\xi \in \overline{\Delta}_{n/m}$, the point $\zeta = e^{2\pi i\xi}$ satisfies*

$$\text{(3.2)} \qquad \mathrm{dist}(\zeta, \overline{\Delta^- \cup \Delta^+}) \geq \mu|\zeta - \lambda|^2 \quad \textit{and} \quad |\zeta - \lambda| \geq \mu\, m^{-\tau}.$$

PROOF. Let $\gamma > 0$ and $y \in \mathrm{DC}(\gamma,\tau)$ such that $\lambda = e^{2\pi i y}$. We choose j large enough to ensure $\gamma_j \leq \frac{1}{4}\gamma \min_{m\geq 1}\{m^{1-\tau}\, e^{\alpha m}\}$. According to Definition 3.5,

$$\forall n/m \in \mathbb{Q}, \quad |y - n/m| \geq \gamma_j\, \frac{e^{-\alpha m}}{m} = \frac{\psi_j(m)}{m},$$

hence $y \in C_{\psi_j} \pmod{\mathbb{Z}}$ and $\lambda \in K_j$ by (2.5).

Let us define the function $f(X) = 2\kappa\gamma_j \exp(-c|X|^{-1/\tau})$, with $c = \alpha(\frac{\gamma}{2})^{1/\tau}$. We can use Lemma 2.4 and Proposition 2.2 to show that

$$\mathcal{K}_f = \{\xi \in \mathbb{C} \mid |\Im\mathrm{m}\,\xi| \geq f(\Re\mathrm{e}(\xi - y))\} \subset \tilde{C}_{\psi_j,\kappa},$$

where $\tilde{C}_{\psi_j,\kappa}$ denotes the lift of $C_{\psi_j,\kappa}$ in $\mathbb{C}$. Indeed, if $\xi \in \mathbb{C} \setminus \tilde{C}_{\psi_j,\kappa}$, there exists $n/m \in \mathbb{Q}_\psi$ such that $\xi \in \Delta_{n/m} \pmod{\mathbb{Z}}$; according to Proposition 2.2 (2),

$$|\Re\mathrm{e}(\xi - n/m)| < 2\gamma_j\, \frac{e^{-\alpha m}}{m} \quad \text{and} \quad |\Im\mathrm{m}\,\xi| \leq \tfrac{1}{2}\kappa(x'_{n/m} - x_{n/m}) < 2\kappa\gamma_j\, e^{-\alpha m};$$

but $X = \Re\mathrm{e}(\xi - y)$ satisfies $|X| \geq |y - n/m| - |\Re\mathrm{e}(\xi - n/m)| \geq \frac{1}{2}\gamma m^{-\tau}$, hence $|\Im\mathrm{m}\,\xi| < f(X)$.

Since $\mathcal{K}_f$ has a contact of infinite order with $\mathbb{R}$ at y, we obtain $\Delta^- \cup \Delta^+ \subset \exp(2\pi i\, C_{\psi_j,\kappa})$ by taking the radius of these disks small enough. Reducing this radius if necessary, we make them contained in the annulus $\{e^{-2\pi d} \leq |q| \leq e^{2\pi d}\}$ and thus in K_j.

Finally, by compactness, it is sufficient to prove (3.2) for $\zeta = e^{2\pi i \xi}$ close to λ. On the one hand, the estimate

$$\mathrm{dist}(\zeta, \overline{\Delta^- \cup \Delta^+}) \underset{\zeta \to \lambda}{\sim} \mathrm{const}\ |\zeta - \lambda|^2$$

follows from the fact that, for all $\zeta \in \mathbb{C} \setminus \exp(2\pi i\, \mathcal{K}_f)$, $|\zeta| = 1 + \tilde{f}(\zeta - \lambda)$, where the function $\tilde{f}(X)$ is exponentially small for small $|X|$. On the other hand, $|\zeta - \lambda| \geq \mathrm{const}\ |\xi - y| \geq \mathrm{const}\ m^{-\tau}$ if $\xi \in \overline{\Delta}_{n/m}$, according to the previous computation. □

REMARK 3.3. The exponent 2 in the right-hand side of the first inequality of (3.2) corresponds to the order of contact of the disks $\Delta^\pm$ in which we ask for asymptotic expansions. But the proof of Theorem 3.3 which follows would be valid with any other exponent as well. This means that a monogenic function of $\mathcal{M}((K_j), B)$ admits a Gevrey-τ asymptotic expansion at λ in compacts with arbitrarily high order of contact at λ, not only disks.

PROOF OF THEOREM 3.3. Let $\tau \geq 2$, $\lambda \in \underline{\mathrm{DC}}_\tau$ and $f \in \mathcal{M}((K_j), B)$. Let μ, j, $\Delta^\pm$ as in Lemma 3.2. We proceed as in the proof of Theorem 2.3: the connected components of $\mathbb{C} \setminus K_j$ are of the form $U_\ell = \exp(2\pi i\, \Delta_{n_\ell/m_\ell})$ with $n_\ell/m_\ell \in \mathbb{Q}_{\psi_j}$, except for $\ell = 1$ or $\ell = 1, 2$. We recall that $\mathrm{length}(\Delta_{n/m}) \leq \mathrm{const}\ \frac{e^{-\alpha m}}{m}$. Formula (2.15) leads us to define the coefficients

$$a_k = \frac{f^{(k)}(\lambda)}{k!} = \frac{1}{2\pi i} \sum_{\ell \geq 1} \int_{\partial U_\ell} \frac{f(\zeta)}{(\zeta - \lambda)^{k+1}}\, d\zeta, \qquad k \geq 0.$$

Cauchy's formula (extended to monogenic functions) applies for $q \in \overline{\Delta^- \cup \Delta^+}$:

$$f(q) = \frac{1}{2\pi i} \sum_{\ell \geq 1} \int_{\partial U_\ell} \frac{f(\zeta)}{\zeta - q}\, d\zeta.$$

Using the identity

$$\frac{1}{\zeta - q} = \sum_{k=0}^{N-1} \frac{(q - \lambda)^k}{(\zeta - \lambda)^{k+1}} + \frac{(q - \lambda)^N}{(\zeta - \lambda)^N (\zeta - q)},$$

we find that

$$f(q) - \sum_{k=0}^{N-1} a_k (q - \lambda)^k = \frac{1}{2\pi i} \sum_{\ell \geq 1} \int_{\partial U_\ell} \frac{f(\zeta)}{(\zeta - \lambda)^N} \frac{(q - \lambda)^N}{\zeta - q}\, d\zeta.$$

We now use (3.2) to bound one by one the contributions of the components $U_\ell = \exp(2\pi i\, \Delta_{n_\ell/m_\ell})$, noticing that, by compactness, such inequalities hold for the exceptional components as well provided that μ is small enough: if $\zeta \in \partial U_\ell$,

$|\zeta - q| \geq \mu|\zeta - \lambda|^2$ and $|\zeta - \lambda| \geq \mu\, m_\ell^{-\tau}$ (extending the definition of m_ℓ by the value 1 for the exceptional components), hence

$$(3.3)\quad \|f(q) - \sum_{k=0}^{N-1} a_k(q-\lambda)^k\| \leq \frac{\text{const}}{\mu^{N+3}} \sum_{\ell\geq 1} \frac{e^{-\alpha m_\ell}}{m_\ell} m_\ell^{(N+2)\tau} \leq \frac{\text{const}}{\mu^{N+3}} \Phi((N+2)\tau),$$

with $\Phi(X) = \sum_{m\geq 1} m^X\, e^{-\alpha m}$. Comparing the sum $\Phi(X)$ and the integral

$$\int_0^{+\infty} m^X\, e^{-\alpha m}\, dm = \alpha^{-X-1}\Gamma(X+1),$$

we obtain $\Phi(X) \leq \alpha^{-X-1}(\Gamma(X+1) + 2\alpha\, X^X\, e^{-X})$ and the Stirling formula yields the result. □

3.3. Borel transform at quadratic irrationals for the fundamental solution

If $\lambda \in \underline{\mathrm{DC}}_\tau$ for some $\tau \geq 2$, according to Theorem 3.3 the solutions of the cohomological equation are contained in the corresponding Gevrey class, which is not quasianalytic at λ. But would it be possible for them to be contained in some smaller, quasianalytic Carleman class? We now show that the answer is *negative* if $\tau = 2$ and λ belongs to a subset QI of $\underline{\mathrm{DC}}_2$.

DEFINITION 3.6. For any point $\lambda = e^{2\pi i\alpha}$ in $\underline{\mathrm{DC}}_2$ (say with $\alpha \in]0,1[$), we define the *Lagrange spectral constants* $\nu_\pm(\lambda) > 0$ by

$$\frac{1}{\nu_-(\lambda)} = -\liminf_{(D,N)\in\mathbb{N}^*\times\mathbb{Z}} D^{-2}\Big(\frac{N}{D} - \alpha\Big)^{-1}, \qquad \frac{1}{\nu_+(\lambda)} = \limsup_{(D,N)\in\mathbb{N}^*\times\mathbb{Z}} D^{-2}\Big(\frac{N}{D} - \alpha\Big)^{-1}.$$

We will use the notation $\kappa_\pm(\lambda) = (\nu_\pm(\lambda))^{1/2}$ too.

The previous definition is clearer when introducing the sets (which depend on α)

$$\mathcal{E}^- = \{\,(D,N) \in \mathbb{N}^*\times\mathbb{Z} \mid N/D < \alpha\,\} \quad \text{and} \quad \mathcal{E}^+ = \{\,(D,N) \in \mathbb{N}^*\times\mathbb{Z} \mid N/D > \alpha\,\},$$

since we can write

$$\nu_\pm(\lambda) = \kappa_\pm(\lambda)^2 = \liminf_{(D,N)\in\mathcal{E}^\pm} D^2\left|\frac{N}{D} - \alpha\right|.$$

DEFINITION 3.7. We define QI to be the subset of $\underline{\mathrm{DC}}_2$ consisting of all $\lambda = e^{2\pi i\alpha}$ with α quadratic irrational, i.e. $\alpha \in \mathbb{R}\setminus\mathbb{Q}$ algebraic of degree 2.

The *Lagrange spectrum* can be defined as the set of all numbers of the form $\nu(\lambda) = \min\{\nu_-(\lambda), \nu_+(\lambda)\}$ for some $\lambda \in \underline{\mathrm{DC}}_2$ (i.e. it is the set of the numbers $\nu(\lambda) = \liminf D^2\left|\frac{N}{D} - \alpha\right|$ for $\lambda \in \underline{\mathrm{DC}}_2$), but here we need an asymmetric version of it because we will separate the rational approximations of α by the left from its rational approximations by the right. The inclusion $\mathrm{QI} \subset \underline{\mathrm{DC}}_2$ (which is tacitly assumed in Definition 3.7) follows e.g. from Lagrange's theorem which is recalled in the appendix (Theorem A3.1). We will need the restriction $\lambda \in \mathrm{QI}$ because of the following lemma, which is an arithmetical result about the way the quantities $D^2\left|\frac{N}{D} - \alpha\right|$ approach $\nu_\pm(\lambda)$, and for which we do not know of any analogue when $\lambda \in \underline{\mathrm{DC}}_2 \setminus \mathrm{QI}$.

LEMMA 3.3. *Let $\lambda = e^{2\pi i\alpha}$ with $\alpha \in]0,1[$ irrational and algebraic of degree 2. For each of the above sets $\mathcal{E}^+$ and $\mathcal{E}^-$, there exist a partition*

$$\mathcal{E}^\pm = \mathcal{F}^\pm \cup \mathcal{E}^\pm_* \cup \mathcal{A}^\pm$$

and a number $\kappa'_\pm > \kappa_\pm(\lambda)$ such that:
- *the set $\mathcal{F}^\pm$ is finite;*
- *for all $(D,N) \in \mathcal{E}^\pm_*$, $D^2\left|\frac{N}{D} - \alpha\right| \geq (\kappa'_\pm)^2$;*
- *the set $\mathcal{A}^\pm$ can be written*

$$\mathcal{A}^\pm = \{\, (D_p^\pm, N_p^\pm),\ p \geq 0\,\}$$

with $\{D_p^\pm\}$ increasing sequence of $\mathbb{N}^*$, $(D_p^\pm)^2\left|\frac{N_p^\pm}{D_p^\pm} - \alpha\right| = \nu_\pm(\lambda) + O(\frac{1}{(D_p^\pm)^2})$ and $\sum (D_p^\pm)^{-1/2} < \infty$. Moreover the sequence $\{D_{p+1}^\pm / D_p^\pm\}$ is bounded.

In the case of the golden mean $\alpha = \frac{1+\sqrt{5}}{2}$, one may check that $\nu_+(\lambda) = \nu_-(\lambda) = \frac{1}{\sqrt{5}}$. But for $\alpha = \sqrt{3}$, one finds $\nu_-(\lambda) = \frac{1}{\sqrt{3}} > \nu_+(\lambda) = \frac{1}{2\sqrt{3}}$. In both examples one can take the even convergents for the sequence $\{\frac{N_p^+}{D_p^+}\}$ and the odd convergents for the sequence $\{\frac{N_p^-}{D_p^-}\}$. In general $\{\frac{N_p^\pm}{D_p^\pm}\}$ is a subsequence of the sequence of odd/even convergents of α.

The proof of Lemma 3.3 is given in Appendix A.4, since it is purely arithmetical. It is the only place where we use the hypothesis $\lambda \in \mathrm{QI}$.

THEOREM 3.4 (Non-quasianalyticity and sharpness of Gevrey-2 asymptotics for quadratic irrationals). *Let $\lambda \in \mathrm{QI}$ and $r \in]0,1[$. We know by Theorem 3.3 that $f_\delta \in \mathcal{G}_2(\lambda, B_r)$, thus we may consider its asymptotic expansion at λ:*

$$\tilde{f} = J_\lambda(f_\delta) = \sum_{n\geq 0} F_n Q^n \in B_r[[Q]]_2,$$

and the formal Borel transform of $Q^{1/2}\tilde{f}(Q)$ with respect to $Q^{1/2}$:

$$\hat{F}(\xi, z) = \sum_{n\geq 0} F_n(z)\frac{\xi^{2n}}{(2n)!} \in \mathbb{C}\{\xi, z\}.$$

(a) The holomorphic germ $\hat{F}$ extends analytically to the set $\{\, (\xi,z) \in \mathbb{C}\times\mathbb{D}_r \mid \xi \in \mathrm{REC}_\lambda(z)\,\}$ where, for each $z \in \mathbb{D}_r$, the rectangle $\mathrm{REC}_\lambda(z)$ is defined as the set of the complex numbers ξ such that

$$\left|\Re\mathrm{e}\big((2\pi i\lambda)^{-1/2}\xi\big)\right| < \kappa_+(\lambda)\log\frac{1}{|z|} \quad \text{and} \quad \left|\Im\mathrm{m}\big((2\pi i\lambda)^{-1/2}\xi\big)\right| < \kappa_-(\lambda)\log\frac{1}{|z|}$$

(see Figure 4).

(b) For each $z \in \mathbb{D}_r$, $\partial(\mathrm{REC}_\lambda(z))$ is a natural boundary for the analytic function $\xi \mapsto \hat{F}(\xi, z)$.

(c) Suppose that $\{M_n\}$ is a non-decreasing sequence of positive numbers such that

$$f_\delta \in \mathcal{C}^-(\lambda, \{M_n\}, B_r) \quad or \quad f_\delta \in \mathcal{C}^+(\lambda, \{M_n\}, B_r).$$

Necessarily $\mathcal{C}^\pm(\lambda, \{M_n\}, B_r)$ is not quasianalytic at λ.

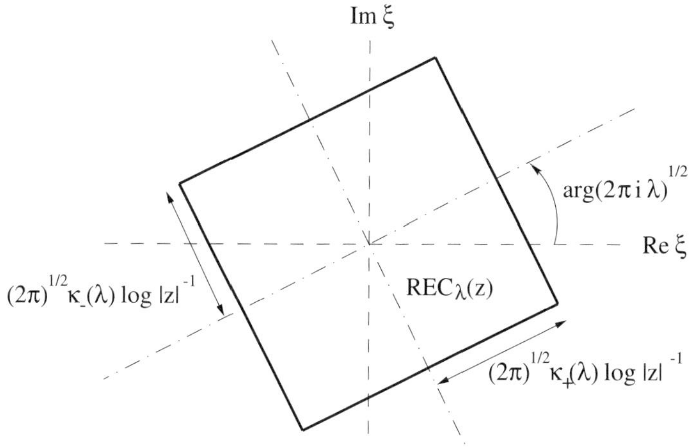

FIGURE 4

The assumption that $\{M_n\}$ be non-decreasing seems only technical, but we were not able to get rid of it. With that restriction the spaces of solutions $\{f_g,\ g \in B_{r'}\}$ with $r' > r$ and *a fortiori* the spaces of monogenic functions $\mathcal{M}(\{C_j\}, B_{r'})$ are contained in none of the quasianalytic Carleman classes at λ that we have defined in Section 3.1.

Note that this theorem holds for the fundamental solution of the cohomological equation, because of its very specific features, but we claim no such result for a general Borel-Wolff-Denjoy series with poles in $\mathcal{R}$ nor for any class of monogenic functions.

We will obtain that theorem itself as a consequence of a more precise result. In the statement of this result, we will make use of the variables $h = \frac{1}{2\pi i} \log \frac{q}{\lambda}$ and $s = \log z$ rather than $Q = q - \lambda$ and z. Since we are dealing with functions of B_r, the variable s will move in the half-plane $\{\, \Re\mathrm{e}\, s < \log r < 0 \,\}$ and these functions decrease at least like $e^{\Re\mathrm{e}\, s}$ when $\Re\mathrm{e}\, s$ tends to $-\infty$.

THEOREM 3.5 (Borel transform of order 2 at quadratic irrationals). *Let $\lambda \in$ QI and $r \in \,]0,1[$. One can give a decomposition of the fundamental solution*

$$f_\delta(\lambda\, e^{2\pi i h}, e^s) = f_\delta(\lambda, e^s) + \frac{1}{2\pi i}\big(\chi^+(h,s) + \chi^-(h,s)\big)$$

satisfying the following properties:

(a) the function $\chi^\pm$ is analytic for $\Re\mathrm{e}\, s < \log r$ and $h \in \mathbb{C} \setminus \mathbb{R}^\pm$, with

$$\chi^+(h,s) = h^{1/2} \int_0^{+i\infty} \hat{\psi}^+(\zeta, s)\, e^{-\zeta h^{-1/2}}\, d\zeta,$$

$$\chi^-(h,s) = h^{1/2} \int_0^{+\infty} \hat{\psi}^-(\zeta, s)\, e^{-\zeta h^{-1/2}}\, d\zeta,$$

the Borel transform $\hat{\psi}^\pm$ being analytic in

$$\{\, (\zeta, s) \mid \Re\mathrm{e}\, s < \log r \text{ and } |\Re\mathrm{e}\, \zeta| < \kappa_+(\lambda)(-\Re\mathrm{e}\, s) \,\} \text{ for } \hat{\psi}^+,$$

$$\{\, (\zeta, s) \mid \Re\mathrm{e}\, s < \log r \text{ and } |\Im\mathrm{m}\, \zeta| < \kappa_-(\lambda)(-\Re\mathrm{e}\, s) \,\} \text{ for } \hat{\psi}^-,$$

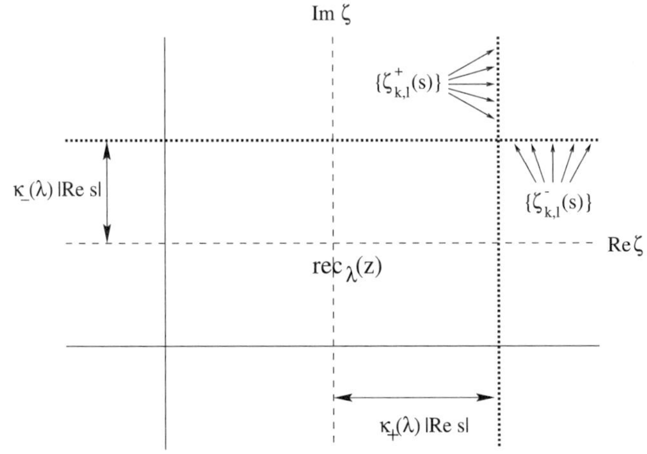

FIGURE 5

and, for each s, even with respect to ζ and bounded in any substrip of the form

$$\{\,|\Re e\,\zeta| \le \text{const}\,\}\ \text{for}\ \hat{\psi}^+, \quad \text{or}\ \{\,|\Im m\,\zeta| \le \text{const}\,\}\ \text{for}\ \hat{\psi}^-,$$

where const *denotes some number* $< \kappa_\pm(\lambda)(-\Re e\, s)$*;*

(b) for each s, the Borel transform $\zeta \mapsto \hat{\psi}^\pm(\zeta, s)$ has a dense set of singular points on the boundary of its strip of definition; more precisely, if one defines the points

$$\zeta^+_{k,l}(s) = \kappa_+(\lambda)\big(-s+2\pi i(k\alpha+l)\big), \quad \zeta^-_{k,l}(s) = i\kappa_-(\lambda)\big(-s+2\pi i(k\alpha+l)\big), \qquad k,l \in \mathbb{Z},$$

the real part of the function $\hat{\psi}^\pm(\zeta,s)$ tends to $-\infty$ when ζ tends to one of the points $\zeta^\pm_{k,l}(s)$, horizontally from the left for $\hat{\psi}^+$, vertically from below for $\hat{\psi}^-$ (i.e. $\zeta = \zeta^\pm_{k,l}(s) + \xi$, $\xi \to 0$, with $\xi \in \mathbb{R}^-$ for $\hat{\psi}^+$ and $\xi \in i\,\mathbb{R}^-$ for $\hat{\psi}^-$; see Figure 5).

(c) for each real $s < \log r$ there exists a positive integer j_0 and a non-decreasing sequence of positive numbers $\{\delta_j\}_{j \ge j_0}$ such that

$$\sum_{j \ge j_0} (\delta_j)^{-3/4} < +\infty \qquad \text{and} \qquad \forall j \ge j_0, \quad |\chi_{2j-1}(s)| \ge (\delta_j)^{2j-1},$$

with the following notation for the Taylor series of $\hat{\psi} = \hat{\psi}^+ + \hat{\psi}^-$:

$$\hat{\psi}(\zeta, s) = \sum_{n \ge 0} \chi_{n+1}(s) \frac{\zeta^{2n}}{(2n)!}. \tag{3.4}$$

REMARK 3.4. The function $\zeta \mapsto \hat{\psi}^\pm(\zeta, s)$ is in fact the *integral* Borel transform of $h \mapsto h^{-1/2}\chi^\pm(h,s)$ with respect to $h^{1/2}$, whereas its Taylor series at $\zeta = 0$ is the *formal* Borel transform with respect to $h^{1/2}$ of the asymptotic expansion of $h^{-1/2}\chi^\pm$ at $h = 0$. In the formulas of Part (a) which indicate how to recover $\chi^\pm$ from $\hat{\psi}^\pm$ by Laplace transform, there is an implicit choice of determination of $h^{1/2}$: for χ^+ one chooses the determination which is holomorphic in $\mathbb{C} \setminus \mathbb{R}^+$ and has always positive imaginary part, while for χ^- one chooses the determination which

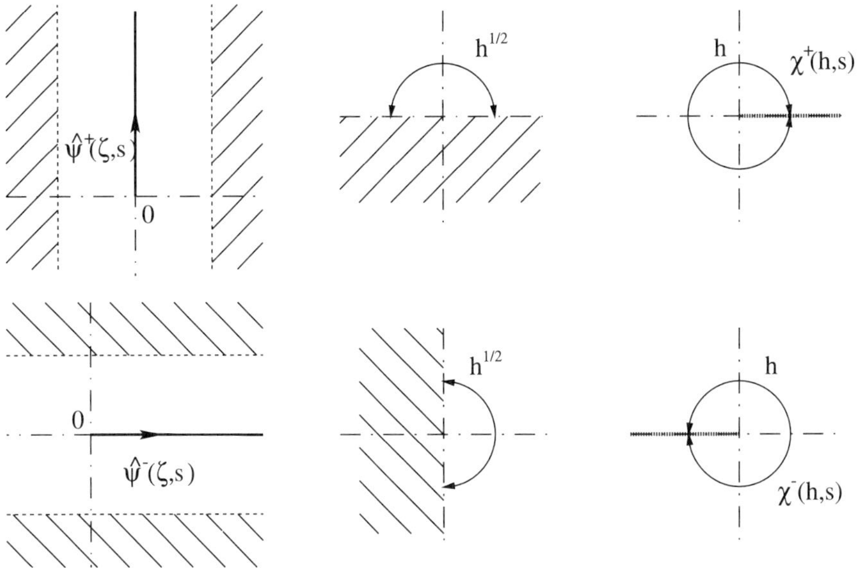

FIGURE 6

is holomorphic in $\mathbb{C} \setminus \mathbb{R}^-$ and has always positive real part (in order to ensure the decrease of $|e^{-\zeta h^{-1/2}}|$); see Figure 6.

One could make the opposite choice as well: since $\hat{\psi}^{\pm}$ is even with respect to ζ, one would simply have to compute the Laplace integral along the opposite ray. Besides, Parts (a) of Theorem 3.4 or 3.5 do not require $\lambda \in \mathrm{QI}$ but only $\lambda \in \underline{\mathrm{DC}}_2$.

3.4. Deduction of Theorem 3.4 from Theorem 3.5

3.4.1. Parts (a) and (b) are an exercise of application of the general theory of which Appendix A.5 gives a brief account. We will relate $\hat{F}(\xi)$ and

$$\hat{\psi}(\zeta) = \hat{\psi}^+(\zeta) + \hat{\psi}^-(\zeta) \tag{3.5}$$

(from now on the variable $z = e^s$ will be understood). Part (a) of Theorem 3.5 implies that $\hat{\psi}$ is analytic in the rectangle $\mathrm{rec}_\lambda(z)$ defined by

$$|\Re\mathrm{e}\,\zeta| < \kappa_+(\lambda)(-\Re\mathrm{e}\, s) \ \text{ and } \ |\Im\mathrm{m}\,\zeta| < \kappa_-(\lambda)(-\Re\mathrm{e}\, s).$$

Let $Q_1 = Q^{1/2}$ and $\tilde{F}(Q_1) = Q_1 \tilde{f}(Q_1^2)$: $\hat{F}(\xi)$ is the formal Borel transform of $\tilde{F}$ with respect to Q_1, which we will indicate by the notation

$$\tilde{F} = \tilde{\mathcal{L}}_{(\xi \to Q_1)} \hat{F}$$

in order to be able to deal with changes of variables in the formal model. By definition of $\tilde{f}$, we have the asymptotic expansion $f_\delta(\lambda + Q_1^2) \sim Q_1^{-1}\tilde{F}(Q_1)$, thus

$$f_\delta(\lambda + Q_1^2) \sim \hat{F}(0) + \tilde{\mathcal{L}}_{(\xi \to Q_1)}(\partial_\xi \hat{F}).$$

On the other hand we can introduce $h_1 = h^{1/2}$. According to Part (a) of Theorem 3.5, $f_\delta(\lambda\, e^{2\pi i h_1^2}) \sim f_\delta(\lambda) + \frac{1}{2\pi i} h_1 \tilde{\mathcal{L}}_{(\zeta \to h_1)} \hat{\psi} = f_\delta(\lambda) + \tilde{\mathcal{L}}_{(\zeta \to h_1)}(\frac{1}{2\pi i} * \hat{\psi})$. We deduce that $\hat{F}(0) = f_\delta(\lambda)$, and setting

$$\hat{G}_1 = \frac{1}{2\pi i} * \hat{\psi}, \tag{3.6}$$

we have the identity $\tilde{\mathcal{L}}_{(\xi\to Q_1)}(\partial_\xi \hat{F}) = \tilde{\mathcal{L}}_{(\zeta\to h_1)}\hat{G}_1$ under the change of variable

$$h_1 = \Big[\frac{1}{2\pi i}\log(1+\lambda^{-1}Q_1^2)\Big]^{1/2} = (2\pi i\lambda)^{-1/2}Q_1(1+O(Q_1^2)).$$

This change of variable is the composition of the dilatation $h_1 \mapsto Q_2 = (2\pi i\lambda)^{1/2}h_1$ and of the transformation $Q_1 \mapsto Q_2 = \big[\lambda\log(1+\lambda^{-1}Q_1^2)\big]^{1/2}$. The dilatation is responsible for the passage from $\hat{G}_1$ analytic for $\zeta \in \mathrm{rec}_\lambda(z)$ to a function

$$\hat{G}_2(\xi_2) = (2\pi i\lambda)^{-1/2}\hat{G}_1((2\pi i\lambda)^{-1/2}\xi_2) \tag{3.7}$$

analytic for $\xi_2 \in \mathrm{REC}_\lambda(z) = (2\pi i\lambda)^{1/2}\mathrm{rec}_\lambda(z)$ and such that

$$\tilde{\mathcal{L}}_{(\zeta\to h_1)}\hat{G}_1 = \tilde{\mathcal{L}}_{(\xi_2\to Q_2)}\hat{G}_2.$$

According to Part (b) of Theorem 3.4, $\partial\mathrm{REC}_\lambda(z)$ is a natural boundary for $\hat{G}_2$.

Finally

$$\hat{F} = \hat{F}(0) + 1 * \hat{G},$$

where the function $\hat{G}(\xi)$ is determined from $\hat{G}_2$ by *composition-convolution*: indeed

$$\tilde{\mathcal{L}}_{(\xi\to Q_1)}\hat{G} = \tilde{\mathcal{L}}_{(\xi_2\to Q_2)}\hat{G}_2$$

under a change of variable

$$Q_2^{-1} = Q_1^{-1} + L_{12}(Q_1) \ \Leftrightarrow\ Q_1^{-1} = Q_2^{-1} + L_{21}(Q_2), \qquad L_{12}(X), L_{21}(X) \in X\mathbb{C}\{X\},$$

hence

$$\hat{G} = \hat{G}_2 + \sum_{r\geq 1}\frac{1}{r!}(\hat{L}_{12})^{*r} * \hat{\partial}^r\hat{G}_2, \quad \hat{G}_2 = \hat{G} + \sum_{r\geq 1}\frac{1}{r!}(\hat{L}_{21})^{*r} * \hat{\partial}^r\hat{G}, \tag{3.8}$$

where $\hat{\partial}$ denotes the multiplication by $-\xi$ or $-\xi_2$, Borel counterpart of differentiation with respect to $X_1 = Q_1^{-1}$ or $X_2 = Q_2^{-1}$. Here $\hat{L}_{12}$ and $\hat{L}_{21}$ are entire functions and $\mathrm{REC}_\lambda(z)$ is star-shaped with respect to the origin, hence the above series are uniformly convergent in any compact subset of $\mathrm{REC}_\lambda(z)$. Therefore $\hat{G}$ is holomorphic in $\mathrm{REC}_\lambda(z)$, and if $\partial\mathrm{REC}_\lambda(z)$ were not a natural boundary for $\hat{G}$, neither would it be for $\hat{G}_2$. This proves the statements of Parts (a) and (b) of Theorem 3.4.

3.4.2. As for Part (c), we now suppose that $f_\delta \in \mathcal{C}^\pm(\lambda, \{M_n\}, B_r)$ for some non-decreasing sequence of positive numbers $\{M_n\}$. In particular $\tilde{f} = J_\lambda^\pm(f_\delta) \in B_r[[Q]]_{\{M_n\}}$. Let us fix a real number $s_0 < \log r$ at which all the subsequent s-dependent functions will be evaluated. For instance F_n will denote the value at s_0 of the function $s \mapsto F_n(e^s)$, and we have

$$\forall n \geq 1, \quad |F_n| \leq c_0 c_1^n M_n$$

for some $c_0, c_1 > 0$.

Part (c) of Theorem 3.5 yields a sequence $\{\delta_j\}$ which allows us to bound from below half of the coefficients of

$$\tilde{\chi}(h) = \sum_{n\geq 0} \chi_n h^n,$$

where we use the convention of (3.4) for denoting the Taylor coefficients of $\hat{\psi}$ and we set $\chi_0 = 2\pi i f_\delta(\lambda)$ for conveniency. We have

$$f_\delta(\lambda + Q) \sim \tilde{f}(Q) = \sum_{n\geq 0} F_n Q^n \quad \text{and} \quad f_\delta(\lambda\, e^{2\pi i h}) \sim \frac{1}{2\pi i}\tilde{\chi}(h),$$

therefore $\tilde{\chi}(h) = 2\pi i \tilde{f}(\lambda(e^{2\pi i h} - 1))$ and in particular, for all $n \geq 1$,

$$\chi_n = (2\pi i)^{n+1} \sum_{r=1}^{n} \lambda^r b_{r,n} F_r \qquad \text{with} \quad b_{r,n} = \sum_{n_1+\cdots+n_r=n,\, n_i\geq 1} \frac{1}{n_1! \cdots n_r!}.$$

Since $b_{r,n} < r^n/n!$ and $\{M_n\}$ is non-decreasing, we deduce that

$$|\chi_n| \leq c_0 (2\pi)^{n+1} M_n \frac{n^{n+1}}{n!} (\max\{1, c_1\})^n,$$

and for n large enough, $M_n \geq c_2^n |\chi_n|$ with some $c_2 > 0$.

We are now in a position to apply Theorem 3.1: let $\beta_n = \inf_{n'\geq n} \{ M_{n'}^{1/n'} \}$. For j large enough,

$$M_{2j} \geq M_{2j-1} \geq c_2^{2j-1} (\delta_j)^{2j-1},$$

thus, for j large enough,

$$M_{2j}^{\frac{1}{2j}} \geq c_2^{1-\frac{1}{2j}} (\delta_j)^{1-\frac{1}{2j}} \geq \text{const}\ (\delta_j)^{3/4} \quad \text{and} \quad M_{2j-1}^{\frac{1}{2j-1}} \geq c_2 \delta_j \geq \text{const}\ (\delta_j)^{3/4}.$$

Since the sequence $\{\delta_n\}$ is non-decreasing,

$$\beta_{2j} \geq \text{const}\ (\delta_j)^{3/4}, \quad \beta_{2j-1} \geq \text{const}\ (\delta_j)^{3/4},$$

thus $\sum \frac{1}{\beta_n} < +\infty$.

3.5. Proof of Theorem 3.5

3.5.1. Let $\lambda = e^{2\pi i \alpha} \in \mathrm{QI}$ with $\alpha \in]0,1[$. We first give a formula close to a "decomposition into simple elements" of f_δ with respect to the variable $h = \frac{1}{2\pi i} \log \frac{q}{\lambda}$.

LEMMA 3.4.

$$f_\delta(\lambda\, e^{2\pi i h}, z) = f_\delta(\lambda, z) + \frac{1}{2\pi i} \sum_{D\in\mathbb{N}^*,\, N\in\mathbb{Z}} Z^{-1} \cdot \frac{h}{h - Z} \cdot \frac{z^D}{D}, \qquad \textit{with } Z = \frac{N}{D} - \alpha.$$

Observe that in the above formula we have not tried to group the various terms corresponding to the same Z. Doing so, we would arrive to a formula which looks more like a decomposition into simple elements and is more reminiscent of the decomposition (1.5):

$$f_\delta(\lambda\, e^{2\pi i h}, z) = f_\delta(\lambda, z) + \frac{1}{2\pi i} \sum_{Z\in -\alpha + \mathbb{Q}} Z^{-1} \cdot \frac{h}{h - Z} \cdot \mathcal{L}_{D(Z)}(z),$$

where $D(Z)$ denotes the denominator of $Z+\alpha$.

PROOF OF LEMMA 3.4. We start with the decomposition which is relative to the variable q and which can be written

$$f_\delta(q,z)=\sum_{D\geq 1}\frac{z^D}{D}\sum_{\Lambda\in\mathcal{R}_D}\Big(\frac{q}{\Lambda}-1\Big)^{-1}.$$

Thus

$$f_\delta(\lambda\,e^{2\pi ih},z)=\sum_{D\geq 1}\sum_{0\leq N\leq D-1}\frac{z^D}{D}\big(e^{2\pi i(h+\alpha-\frac{N}{D})}-1\big)^{-1}.$$

We now use the identity

$$\frac{d}{dx}(e^{2\pi ix}-1)^{-1}=\frac{1}{2\pi i}\sum_{M\in\mathbb{Z}}\frac{d}{dx}(x-M)^{-1},$$

which yields

$$\begin{aligned}\frac{d}{dh}[f(\lambda\,e^{2\pi ih},z)]&=\sum_{D\geq 1}\sum_{0\leq N\leq D-1}\frac{1}{2\pi i}\frac{z^D}{D}\sum_{M\in\mathbb{Z}}\frac{d}{dh}\Big(h+\alpha-\frac{N+MD}{D}\Big)^{-1}\\&=\frac{1}{2\pi i}\sum_{D\geq 1}\sum_{N\in\mathbb{Z}}\frac{z^D}{D}\frac{d}{dh}\Big(h+\alpha-\frac{N}{D}\Big)^{-1}\\&=\frac{1}{2\pi i}\sum_{D\geq 1}\sum_{N\in\mathbb{Z}}\frac{z^D}{D}\frac{d}{dh}\Big[Z^{-1}\cdot\frac{h}{h-Z}\Big],\end{aligned}$$

hence the result by integration. □

Using the notations of Lemma 3.3, we introduce functions χ^+ and χ^- defined by

$$\chi^\pm(h,s)=\sum_{(D,N)\in\mathcal{E}^\pm}Z^{-1}\cdot\frac{h}{h-Z}\cdot\frac{e^{Ds}}{D},$$

still with $Z=\frac{N}{D}-\alpha$. Thus, according to Lemma 3.4,

$$f_\delta(\lambda\,e^{2\pi ih},z)=f_\delta(\lambda,z)+\frac{1}{2\pi i}\big(\chi^+(h,s)+\chi^-(h,s)\big),$$

with the change of variable $z=e^s$. Each term

$$\chi_{(D,N)}(h,s)=Z^{-1}\cdot\frac{h}{h-Z}\cdot\frac{e^{Ds}}{D},$$

being analytic at the origin with respect to $h^{1/2}$, may be written as the Laplace integral in any direction of its Borel transform; we find it convenient to let a factor $h^{1/2}$ outside the integral:

$$\chi_{(D,N)}(h,s)=h^{1/2}\int_0^\infty\hat\psi_{(D,N)}(\zeta,s)\,e^{-\zeta h^{-1/2}}\,d\zeta.$$

One computes easily

$$\hat\psi_{(D,N)}(\zeta,s)=-Z^{-2}\cdot\frac{e^{Ds}}{D}\sum_{n\geq 0}Z^{-n}\cdot\frac{\zeta^{2n}}{(2n)!}$$

which is entire and of exponential type in any direction: according to the sign of Z we obtain a hyperbolic or a trigonometric cosine.

3.5.2. Part (a) of Theorem 3.5 will thus derive from the study of the convergence of the series

$$\hat{\psi}^{+}(\zeta, s) = -\sum_{(D,N)\in\mathcal{E}^{+}} Z^{-2}\cosh\big(Z^{-1/2}\zeta\big)\frac{e^{Ds}}{D} \tag{3.11}$$

and

$$\hat{\psi}^{-}(\zeta, s) = -\sum_{(D,N)\in\mathcal{E}^{-}} Z^{-2}\cos\big(|Z|^{-1/2}\zeta\big)\frac{e^{Ds}}{D}. \tag{3.12}$$

Let us consider $\hat{\psi}^{+}$ for instance. It is of course the even part (with respect to ζ) of

$$\Psi^{+}(\zeta, s) = -\sum_{(D,N)\in\mathcal{E}^{+}} D^{-1}Z^{-2}e^{Z^{-1/2}\zeta+Ds} \tag{3.13}$$

Let $\delta > 0$ and $\kappa_0 < \kappa_{+}(\lambda)$: we obtain the uniform convergence of this series for

$$\Re e\,\zeta + \kappa_0\,\Re e\, s \leq -\delta$$

by observing that $\kappa_{+}(\lambda) = \liminf_{(D,N)\in\mathcal{E}^{+}} DZ^{1/2}$. Indeed for almost all $(D, N) \in \mathcal{E}^{+}$ (i.e. all of them except a finite number), $DZ^{1/2} \geq \kappa_0$, therefore $\Re e(Z^{-1/2}\zeta + Ds) \leq -\delta\kappa_0^{-1}D$; and for each $D \geq 1$,

$$\begin{aligned}\sum_{N\in\mathbb{Z},(D,N)\in\mathcal{E}^{+}} Z^{-2} &= D^2\sum_{N>\alpha D}(N-\alpha D)^{-2}\\ &\leq D^2\Big(\operatorname{dist}(\alpha D,\mathbb{Z})^{-2} + \zeta(2)\Big) \leq \text{const } D^4,\end{aligned}$$

hence

$$|\Psi^{+}(\zeta, s)| \leq \text{const}\sum_{D\geq 1} D^4\, e^{-\delta\kappa_0^{-1}D}$$

in that domain. In particular, Ψ^{+} is analytic and bounded in

$$\{\,(\zeta, s) \mid \Re e\, s < \log r \text{ and } \Re e\,\zeta < \kappa_0(-\Re e\, s) - \delta\,\}$$

for all $\delta > 0$ and $\kappa_0 < \kappa_{+}(\lambda)$, and this is enough to establish the analyticity of $\hat{\psi}^{+}(\zeta, s) = \frac{1}{2}(\Psi^{+}(\zeta, s) + \Psi^{+}(-\zeta, s))$ in

$$\{\,(\zeta, s) \mid \Re e\, s < \log r \text{ and } |\Re e\,\zeta| < \kappa_{+}(\lambda)(-\Re e\, s)\,\}.$$

The same analysis can be performed on $\hat{\psi}^{-}$ which is the even part of

$$\Psi^{-}(\zeta, s) = -\sum_{(D,N)\in\mathcal{E}^{-}} D^{-1}Z^{-2}e^{-|Z|^{-1/2}i\zeta+Ds}, \tag{3.14}$$

but the factor $-i$ in front of ζ in the exponentials is responsible for a rotation of $\pi/2$ of the whole picture.

3.5.3. Lemma 3.3 will be useful in the proof of Part (b) of Theorem 3.5. The partition of $\mathcal{E}^+$ yields a decomposition of Ψ^+:

$$\Psi^+ = \Psi^+_{\mathcal{F}^+} + \Psi^+_{\mathcal{E}^+_*} + \Psi^+_{\mathcal{A}^+}, \quad \text{with } \Psi^+_{\mathcal{B}}(\zeta, s) = -\sum_{(D,N)\in\mathcal{B}} D^{-1} Z^{-2} e^{Z^{-1/2}\zeta + Ds}.$$

Because of the properties of $\mathcal{F}^+$ and $\mathcal{E}^+_*$, the function $\Psi^+_{\mathcal{F}^+} + \Psi^+_{\mathcal{E}^+_*}$ is analytic in a domain

$$\{ (\zeta, s) \mid \Re e\, s < \log r \text{ and } \Re e\, \zeta < \kappa'_+(-\Re e\, s) \}$$

which is larger than the domain of analyticity that we just obtained for Ψ^+, as one can see by the same arguments as above.

Let us fix $s \in \mathbb{C}$ with $\Re e\, s < \log r$ and let us consider a point $\zeta^+_{k,l}(s)$. When ζ tends to $\zeta^+_{k,l}(s)$, the function $\Psi^+_{\mathcal{F}^+} + \Psi^+_{\mathcal{E}^+_*}$ tends to its value at $(\zeta^+_{k,l}(s), s)$, thus its real part remains finite and we now focus on the third term, $\Psi^+_{\mathcal{A}^+}$. According to Lemma 3.3, we can write

$$\Psi^+_{\mathcal{A}^+}(\zeta, s) = -\sum_{p\geq 0} c_p (D^+_p)^3\, e^{(Z^+_p)^{-1/2}\zeta + D^+_p s},$$

with $Z^+_p = \frac{N^+_p}{D^+_p} - \alpha$ and $c_p = (Z^+_p)^{-2}(D^+_p)^{-4}$. Moreover we can study the asymptotic behaviour with respect to p of these quantities:

$$Z^+_p = (\nu_+(\lambda) + \rho^+_p)(D^+_p)^{-2}$$

and $\lim \rho^+_p = 0$, thus $\lim c_p = \nu_+(\lambda)^{-2}$. Let us introduce

$$\sigma_p = (Z^+_p)^{-1/2} - \kappa_+(\lambda)^{-1} D^+_p \sim -\frac{1}{2}\kappa_+(\lambda)^{-3} D^+_p \rho^+_p.$$

We know that $\lim \sigma_p = 0$, and we can define a function

$$\Phi^+(X, \zeta) = \sum_{p\geq 0} c_p (D^+_p)^3 X^{D^+_p}\, e^{\sigma_p \zeta}$$

such that

$$\Psi^+_{\mathcal{A}^+}(\zeta, s) = -\Phi^+(e^{\kappa_+(\lambda)^{-1}\zeta + s}, \zeta).$$

According to the definition of $\zeta^+_{k,l}(s)$, when ζ tends to $\zeta^+_{k,l}(s)$ horizontally by the left, the new variable $X = e^{\kappa_+(\lambda)^{-1}\zeta + s}$ tends to $e^{2\pi i(k\alpha + l)} = \lambda^k$ along the ray $]0, \lambda^k[$. Moreover, since $\operatorname{dist}(k\alpha D^+_p, \mathbb{Z})$ tends to 0 as p tends to infinity, we have $\lim \lambda^{kD^+_p} = 1$. We are in a position to apply the following elementary result:

LEMMA 3.5. *Let $\Phi(X, \zeta) = \sum_{p\geq 0} c_p d_p^3 e^{\sigma_p \zeta} X^{d_p}$. Assume that the σ_p and c_p are real numbers, with $\lim_{p\to\infty} \sigma_p = 0$ and c_p bounded from above and from below by some positive constants, and that $\{d_p\}$ is an increasing sequence of integers such that $\lim_{p\to\infty} \lambda^{kd_p} = 1$. Let $\mathcal{K}$ be a compact subset of $\mathbb{C}$.*

- *The series which defines Φ converges uniformly in $\mathcal{K}_0 \times \mathcal{K}$ for any compact subset $\mathcal{K}_0$ of $\mathbb{D}$.*
- *The function $\Re e\, \Phi(X, \zeta)$ tends to $+\infty$ as X tends to λ^k along the ray $]0, \lambda^k[$, uniformly with respect to $\zeta \in \mathcal{K}$.*

PROOF OF LEMMA 3.5. The convergence of the series is obvious. Let $M > 0$. The quantity $e^{\sigma_p\zeta}\lambda^{kd_p}$ tends to 1 as p tends to infinity uniformly with respect to ζ, thus we can chose p_0 large enough so that, for all $\zeta \in \mathcal{K}$,

$$\Re\mathrm{e}\sum_{p=0}^{p_0} c_p d_p^3\, e^{\sigma_p\zeta}\lambda^{kd_p} \geq 2M \quad \text{and} \quad \forall p \geq p_0,\ \Re\mathrm{e}(e^{\sigma_p\zeta}\lambda^{kd_p}) \geq \frac{1}{2}.$$

Let $\delta > 0$, small enough so that, for all $\zeta \in \mathcal{K}$ and $X \in \mathbb{C}$,

$$|X - \lambda^k| \leq \delta \Rightarrow \left|\sum_{p=0}^{p_0} c_p d_p^3\, e^{\sigma_p\zeta}(X^{d_p} - \lambda^{kd_p})\right| \leq M.$$

We see that, if $\zeta \in \mathcal{K}$ and $X \in]0, \lambda^k[$ with $|X - \lambda^k| \leq \delta$,

$$\begin{aligned}\Re\mathrm{e}\,\Phi(X,\zeta) = \Re\mathrm{e}\sum_{p=0}^{p_0} c_p d_p^3\, e^{\sigma_p\zeta}(X^{d_p} - \lambda^{kd_p}) + \Re\mathrm{e}\sum_{p=0}^{p_0} c_p d_p^3\, e^{\sigma_p\zeta}\lambda^{kd_p}\\ + \Re\mathrm{e}\sum_{p>p_0} c_p d_p^3\, e^{\sigma_p\zeta}X^{d_p} < -M + 2M\end{aligned}$$

(the third term is positive since $X = t\lambda^k$ with $t \in]0,1[$ and $\Re\mathrm{e}(e^{\sigma_p\zeta}\lambda^{kd_p}) > 0$). □

Let us continue the proof of Theorem 3.5. We have obtained that $\Re\mathrm{e}\,\Psi^+_{\mathcal{A}^+}$ and thus $\Re\mathrm{e}\,\Psi^+$ tend to $-\infty$ as ζ tends to $\zeta^+_{k,l}(s)$ horizontally by the left. This allows us to reach the desired conclusion for $\hat{\psi}^+$. The previous work is easily adapted to the case of $\hat{\psi}^-$, with the introduction of

$$\Psi^-_{\mathcal{A}^-}(\zeta, s) = -\sum_{p\geq 0} c_p (D_p^-)^3\, e^{-(Z_p^-)^{-1/2}i\zeta + D_p^- s},$$

(with real numbers c_p and σ_p associated to $\mathcal{A}^-$) and

$$\Phi^-(X,\xi) = \sum_{p\geq 0} c_p (D_p^-)^3 X^{D_p^-}\, e^{\sigma_p\xi},$$

but this time the correspondence is $\Psi^-_{\mathcal{A}^-}(\zeta, s) = \Phi^-(e^{-\kappa_-(\lambda)^{-1}i\zeta+s}, -i\zeta)$. This ends the proof of Part (b) of Theorem 3.5.

3.5.4. We now come to Part (c). Let us fix $s < \log r$ and $z = e^s$ (thus $z \in]0,1[$). We recall the notation

$$\hat{\psi} = \hat{\psi}^+ + \hat{\psi}^- = \sum_{n\geq 0} \chi_{n+1}(s)\frac{\zeta^{2n}}{(2n)!}.$$

Our aim is to bound from below half of the coefficients of that series. According to the formulas (3.11)–(3.12),

$$\forall n \geq 0, \quad -\chi_{n+1}(s) = \sum_{(D,N)\in\mathcal{E}^+} z^D D^{-1} Z^{-n-2} + (-1)^n \sum_{(D,N)\in\mathcal{E}^-} z^D D^{-1}|Z|^{-n-2},$$

with the usual notation $Z = \frac{N}{D} - \alpha$. Let us choose $\varepsilon \in \{+,-\}$ so that $\kappa_\varepsilon(\lambda) \leq \kappa_{-\varepsilon}(\lambda)$. When we restrict ourselves to even n, only positive quantities appear in the right-hand side of the above equation, thus we obtain a lower bound for the left-hand side by retaining only the terms which correspond to $(D, N) \in \mathcal{E}^{\varepsilon}$:

$$\forall j \geq 1, \quad -\chi_{2j-1}(s) > \sum_{p\geq 0} z^{D_p^{\varepsilon}} (D_p^{\varepsilon})^{-1} |Z_p^{\varepsilon}|^{-2j}.$$

According to Lemma 3.3, $|Z_p^{\varepsilon}| = (\nu_{\varepsilon}(\lambda) + \rho_p^{\varepsilon})(D_p^{\varepsilon})^{-2}$ and ρ_p^{ε} tends to 0 as p tends to infinity, thus we can fix p_0 large enough and $c = \frac{3}{2}\nu_{\varepsilon}(\lambda)$ so that

$$\forall p \geq p_0, \quad |Z_p^{\varepsilon}| \leq c(D_p^{\varepsilon})^{-2}.$$

For $j \geq D_{p_0}^{\varepsilon}/4$, we define

$$E_j = \max_{D_p^{\varepsilon} \leq 4j} \{D_p^{\varepsilon}\}.$$

Thus $E_j \leq 4j$, and since $E_j \in \{ D_p^{\varepsilon},\, p \geq p_0 \}$, we can choose to retain only the corresponding contribution in the previous sum:

$$-\chi_{2j-1}(s) > z^{E_j} c^{-2j} E_j^{4j-1} > (c^{-1} z^2)^{2j} E_j^{4j-2},$$

and for j large enough,

$$|\chi_{2j-1}(s)|^{\frac{1}{2j-1}} > \delta_j := \frac{1}{2} c^{-1} z^2 E_j^2.$$

The sequence $\{\delta_j\}$ that we just defined is obviously non-decreasing and there remains only to check that $\sum \delta_j^{-3/4} < +\infty$, i.e. that $\sum E_j^{-3/2} < +\infty$. We observe that $E_j = F_{4j}$ with

$$\forall m \geq D_{p_0}^{\varepsilon}, \quad F_m = \max_{D_p^{\varepsilon} \leq m} \{D_p^{\varepsilon}\},$$

i.e.

$$F_{D_{p_0}^{\varepsilon}} = F_{D_{p_0}^{\varepsilon}+1} = \ldots = F_{D_{p_0+1}^{\varepsilon}-1} = D_{p_0}^{\varepsilon},$$
$$F_{D_{p_0+1}^{\varepsilon}} = F_{D_{p_0+1}^{\varepsilon}+1} = \ldots = F_{D_{p_0+2}^{\varepsilon}-1} = D_{p_0+1}^{\varepsilon},$$

and so on. Hence, for $P > p_0$,

$$\sum_{m=D_{p_0}^{\varepsilon}}^{D_{P+1}^{\varepsilon}-1} F_m^{-3/2} = \sum_{p=p_0}^{P} (D_p^{\varepsilon})^{-3/2} (D_{p+1}^{\varepsilon} - D_p^{\varepsilon}) \leq \left(-1 + \sup \frac{D_{p+1}^{\varepsilon}}{D_p^{\varepsilon}}\right) \sum_{p=p_0}^{P} (D_p^{\varepsilon})^{-1/2},$$

and the series $\sum F_m^{-3/2}$ and $\sum E_j^{-3/2}$ converge. □

CHAPTER 4

Resummation at Resonances and Constant-Type Points

For a class of monogenic functions (to which the solutions of the cohomological equation belong), we have obtained asymptotic expansions at Diophantine points of the unit circle. Now, restricting ourselves to the subspace of Borel-Wolff-Denjoy series with poles at resonances, we will study asymptotic behaviour at resonances.

Then, we will address the question: Is it possible to recover any solution in a constructive way from its asymptotic expansion at a particular point of $\mathbb{S}^1$? We will provide refined results on Gevrey-1 asymptotics at resonances and Gevrey-2 asymptotics at constant-type points which show that the answer is positive for each of these points. In the latter case there is no contradiction with the non-quasianalyticity of $\mathcal{G}_2(\lambda, B)$ nor with Part (c) of Theorem 3.4, since the question amounts to working in a smaller quasianalytic subspace without demanding it to be a Carleman class.

At resonances a rigid structure appears, which is an elementary case of *resurgence* [E1] in the case of the fundamental solution f_δ. The Borel transform of a given solution $f = f_\delta \odot g$ at a resonance Λ_0 can be completely described, the appropriate Laplace transform then yields the function inside or outside the unit disk, and one can even recover all the other residues $\Lambda \mathcal{L}_{m(\Lambda)} \odot g$ from the singularities of the Borel transform at Λ_0 by computing the *Stokes phenomenon*. In some sense, this means passing from local information (one particular singular point Λ_0) to global information (the whole set of "poles").

For constant-type points, although it is likely that no quasianalytic Carleman class contains the solutions (as is the case for quadratic irrationals), one can still define a quasianalytic space which contains them and in which an adaptation of Borel-Laplace summation process provides constructive quasianalytic continuation, like for resonances.

4.1. Asymptotic expansions at resonances

Recall the formulas (2.10) and (2.11) which, by Theorem 2.2, define $\Sigma_{\mathcal{R}} : \mathcal{S}(r, B) \to \mathcal{M}((K_j), B)$.

THEOREM 4.1. *Let $r \in]0,1[$, B a Banach space and $\Lambda_0 \in \mathcal{R}$. If $a \in \mathcal{S}(r, B)$, the function $q \mapsto (q - \Lambda_0)\big(\Sigma_{\mathcal{R}}(a)\big)(q)$ belongs to $\mathcal{G}_1(\Lambda_0, B)$ and the constant term in its asymptotic expansion $J_{\Lambda_0}\big((q - \Lambda_0)\Sigma_{\mathcal{R}}(a)\big)$ is equal to a_{Λ_0}. In particular, if $0 < r_1 < r_2$, the solution F_{r_1,r_2} belongs to $(q - \Lambda_0)^{-1}\mathcal{G}_1(\Lambda_0, \mathcal{L}(B_{r_1}, B_{r_2}))$ and the constant term in $J_{\Lambda_0}\big((q - \Lambda_0)F_{r_1,r_2}\big)$ is $\Lambda_0 \mathcal{L}_{m(\Lambda_0)} \odot$.*

Therefore the Borel-Wolff-Denjoy series of $\Sigma_{\mathcal{R}}\big(\mathcal{S}(r, B)\big)$ or the solutions of the cohomological equation are contained in quasianalytic spaces $(q - \Lambda_0)^{-1}\mathcal{G}_1(\Lambda_0, B)$.

Moreover Nevanlinna's Theorem ensures the possibility of following the quasianalytic continuation of any such Borel-Wolff-Denjoy series f across $\mathbb{S}^1$ "through Λ_0": the Borel transform of $J_{\Lambda_0}\big((q-\Lambda_0)f\big)$ necessarily belongs to $\hat{\mathcal{N}}(B)$, and the appropriate Laplace transform restores the function on one side or the other of $\mathbb{S}^1$. But much more can be said about the Borel transform in the case of the solutions, as will be shown in Section 4.2.

Unfortunately nothing indicates that such a quasianalytic property could be shared by all the monogenic functions of $\mathcal{M}((K_j),B)$ or $\mathcal{C}^\infty_{hol}((K^*_{A,j}),B)$.

PROOF OF THEOREM 4.1. Let $a \in \mathcal{S}(r,B)$ and

$$F(q) = \sum_{\Lambda\in\mathcal{R},\,\Lambda\neq\Lambda_0} \frac{a_\Lambda}{q-\Lambda} = \big(\Sigma_{\mathcal{R}}(a)\big)(q) - \frac{a_{\Lambda_0}}{q-\Lambda_0}.$$

It is sufficient to prove that $F \in \mathcal{G}_1(\Lambda_0, B)$.

We have $\Lambda_0 = e^{2\pi i\alpha}$ with, $\forall n/m \in \mathbb{Q}\setminus\{\alpha\}$, $|\alpha - n/m| \geq 1/(m(\Lambda_0)|m|)$, and one checks easily the existence of a positive constant γ_1 such that

$$\forall \Lambda \in \mathcal{R}\setminus\{\Lambda_0\}, \quad |\Lambda_0 - \Lambda| \geq \frac{\gamma_1}{m(\Lambda)}. \tag{4.1}$$

Therefore the series

$$A_n = (-1)^n \sum_{\Lambda\in\mathcal{R},\,\Lambda\neq\Lambda_0} \frac{a_\Lambda}{(\Lambda_0-\Lambda)^{n+1}}, \qquad n\in\mathbb{N},$$

are absolutely convergent in B. In fact there exists $c>0$ such that, $\forall n \geq 0$, $\|A_n\| \leq c\,\varphi(n+1)$, where the function φ is defined by

$$\forall n\geq 0, \quad \varphi(n) = \sum_{\Lambda\in\mathcal{R},\,\Lambda\neq\Lambda_0} \frac{r^{m(\Lambda)}}{m(\Lambda)} |\Lambda_0-\Lambda|^{-n}.$$

LEMMA 4.1. *Let $\mathcal{K}$ be a compact subset of $\mathbb{C}$ which intersects $\mathbb{S}^1$ at Λ_0 only, with finite order of contact $\beta > 0$ (i.e. $\exists c > 0$ such that $\forall q\in\mathcal{K}$, $\forall q' \in \mathbb{S}^1$, $|q-q'| \geq c|\Lambda_0 - q'|^\beta$). There exists $c_0 > 0$ such that*

$$\forall N\geq 0,\ \forall q\in\mathcal{K}, \quad \|F(q) - \sum_{0\leq n\leq N-1} A_n(q-\Lambda_0)^n\| \leq c_0\,|q-\Lambda_0|^N\,\varphi(N+\beta).$$

PROOF OF LEMMA 4.1. One computes easily the identity

$$F(q) - \sum_{0\leq n\leq N-1} A_n(q-\Lambda_0)^n = (-1)^N(q-\Lambda_0)^N \sum_{\Lambda\in\mathcal{R},\,\Lambda\neq\Lambda_0} a_\Lambda \frac{(\Lambda_0-\Lambda)^{-N}}{q-\Lambda}.$$

But for $q\in\mathcal{K}$ and $\Lambda\in\mathcal{R}$, $|q-\Lambda| \geq c|\Lambda_0-\Lambda|^\beta$, whereas $\|a_\Lambda\| \leq \text{const}\ \frac{r^{m(\Lambda)}}{m(\Lambda)}$. □

We will use this lemma to finish the proof of Theorem 4.1. Let us check the existence of $c_1 > 0$ such that

$$\forall n\geq 0, \quad \varphi(n) \leq c_1^{n+1} n!.$$

Using Inequality (4.1) we obtain

$$\forall n\geq 0, \quad \varphi(n) \leq \gamma_1^{-n} \sum_{m\geq 1} m^n\, r^m.$$

If we set $r = e^{-s}$ with $s > 0$ and compare the sum $\sum_{m\geq 1} m^n\, e^{-ms}$ and the integral $\int_0^{+\infty} m^n\, e^{-ms}\, dm = s^{-n-1} n!$, we obtain $\sum m^n\, e^{-ms} \leq s^{-n-1}(\Gamma(n+1) + 2s\, n^n\, e^{-n})$; the Stirling formula yields the desired inequality.

We now choose for $\mathcal{K}$ a closed disk $\bar{\Delta}^{\pm}$ contained in $\overline{\mathbb{D}}$ or $\overline{\mathbb{E}}$, then $\beta = 2$ and $F_{|\bar{\Delta}^{\pm}} \in \mathcal{G}_1^{\pm}(\Lambda_0, B)$ with $J^{\pm}_{\Lambda_0}(F_{|\bar{\Delta}^{\pm}}) = \sum A_n Q^n$. □

Notice that, according to the proof of Theorem 3.5, if $a \in \mathcal{S}(r, B)$ the Borel-Wolff-Denjoy series $\Sigma_{\mathcal{R}}(a)$ admits a Gevrey-1 asymptotic expansion at Λ_0 in compact subsets $\mathcal{K}$ with arbitrarily high order of contact at Λ_0.

4.2. Resurgence of the fundamental solution at resonances

4.2.1. We fix in this section a resonant point $\Lambda_0 \in \mathcal{R}$ with $m_0 = m(\Lambda_0)$. We denote by n_0 the integer such that

$$\Lambda_0 = e^{2\pi i n_0/m_0}, \quad 0 \leq n_0 \leq m_0 - 1, \quad (n_0|m_0) = 1.$$

We know by Theorem 4.1 that the function $(q - \Lambda_0) f_\delta$ belongs to $\mathcal{G}_1(\Lambda_0, B_r)$ for all $r \in]0, 1[$, with an asymptotic expansion

$$J_{\Lambda_0}\big((q - \Lambda_0) f_\delta\big) = \sum_{n\geq 0} a_n Q^n, \quad a_0 = \Lambda_0 \mathcal{L}_{m_0}, \quad (\forall n \geq 0)\; a_n \in B_r.$$

According to Theorem 3.2 and Definitions 3.3 and 3.4, the Borel transform

$$\hat{\Phi}^\delta(\xi) = \sum_{n\geq 0} \Lambda_0^{n+1} a_{n+1} \frac{\xi^n}{n!}$$

belongs to $\hat{\mathcal{N}}(B_r)$ for all $r \in]0, 1[$, and f_δ can be recovered from $\hat{\Phi}^\delta$ by the formula $(q - \Lambda_0) f_\delta = \Lambda_0 \mathcal{L}_{m_0} + \mathbb{L}_{\Lambda_0} \hat{\Phi}^\delta$, which can be rephrased as

$$f_\delta\big(\Lambda_0(1 + t)\big) = t^{-1}\mathcal{L}_{m_0} + \Lambda_0^{-1} t^{-1} \Phi^\delta(t), \qquad \Phi^\delta(t) = \int_0^{\pm\infty} \hat{\Phi}^\delta(\xi)\, e^{-\xi/t}\, d\xi.$$

We may consider $\hat{\Phi}^\delta$ as a holomorphic function of two variables as well, by setting $\hat{\Phi}^\delta(\xi, z) = \hat{\Phi}^\delta(\xi)(z)$. Our goal is now to study the analytic continuation with respect to ξ of this Borel transform.

Definition 4.1. For $a \in \mathbb{Z}^*$ and $b \in \mathbb{Z}$, we define the *moving singular point*

$$z \in \mathbb{D}^* \quad \mapsto \quad \xi_{a,b}(z) = \frac{2\pi a}{m_0}\left(-i \log z + \frac{2\pi b}{m_0}\right) \in \mathbb{C},$$

where $\mathbb{D}^* = \mathbb{D} \setminus \{0\}$ and we have chosen some determination of the logarithm once for all (see Figure 7). We also attach to it a complex number:

$$C_{a,b} = -\frac{1}{m_0}\, e^{2\pi i a b n_0'/m_0},$$

where $n_0' + m_0\mathbb{Z}$ is the multiplicative inverse of $n_0 + m_0\mathbb{Z}$ in the ring $\mathbb{Z}/m_0\mathbb{Z}$.

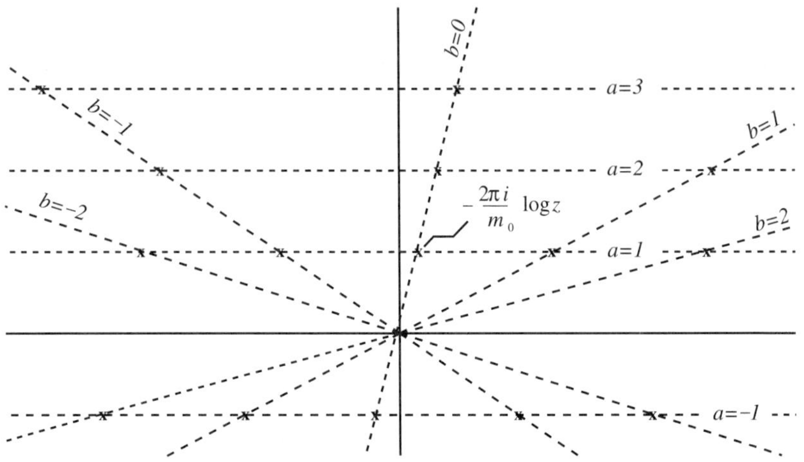

FIGURE 7. The points $\xi_{a,b}(z)$ lie at the intersection of two family of lines parametrized by $a \in \mathbb{Z}^*$ or $b \in \mathbb{Z}$.

THEOREM 4.2 (Resurgence at resonances). *For each $z \in \mathbb{D}^*$, the function $\xi \mapsto \hat{\Phi}^\delta(\xi, z)$ extends analytically to the universal covering*[4] *of $\mathbb{C} \setminus \{\xi_{a,b}(z),\ a \in \mathbb{Z}^*,\ b \in \mathbb{Z}\}$; near a moving singular point $\omega = \xi_{a,b}(z)$ on the main sheet of this Riemann surface, one can write*

$$\hat{\Phi}^\delta(\xi, z) = \Lambda_0 C_{a,b} \left(\frac{e^{-\omega/2}}{\xi - \omega} + \hat{L}_\omega(\xi - \omega) \log(\xi - \omega) \right) + \textit{regular function},$$

where $\hat{L}_\omega$ is an entire function. Moreover, for any $z \in \mathbb{D}^*$ and for any line Δ of $\mathbb{C}$ passing through the origin and avoiding the singular points $\xi_{a,b}(z)$, the function $\hat{\Phi}^\delta(\xi, z)$ has at most exponential growth for $\xi \in \Delta$.

It is even possible to compute the entire functions $\hat{L}_\omega$: they are the Borel transforms of the convergent series $L_\omega(t) = -e^{-\omega/2} + \big(1 + tL(t)\big)e^{-\omega L(t)} = O(t)$, where

$$L(t) = \big(\log(1+t)\big)^{-1} - t^{-1} = \frac{1}{2} + O(t).$$

This theorem will appear as a consequence of Theorem 4.3 below.

In the terminology of resurgence, $(q - \Lambda_0) f_\delta(q)$ would be called a *simple resurgent function* (see Appendix A.5). Theorem 4.2 shows that the index 1 in the Gevrey asymptotics provided by Theorem 4.1 is optimal, since the Borel transform $\hat{\Phi}^\delta$ has finite radius of convergence with respect to ξ for each nonzero z.

REMARK 4.1. There is some analogy between the first line of moving singular points $\xi_{1,b}(z)$ and the points $\zeta^{\pm}_{k,l}(s)$ of Theorem 3.5 (b). Both cases deal with the Borel transform of some Gevrey-τ asymptotic expansion at a point of $\underline{\mathrm{DC}}_\tau$, at $\lambda = e^{2\pi i\alpha} \in \mathrm{QI}$ in Section 3.3 ($\tau = 2$) and at $\Lambda_0 = e^{2\pi i n_0/m_0} \in \mathcal{R}$ here ($\tau = 1$; indeed (4.1) leads us to set $\underline{\mathrm{DC}}_1 = \mathcal{R}$). We have $s = \log z$, but in Section 3.3 we

[4]This simply means that for $\hat{\Phi}^\delta(\xi, z)$ viewed as an analytic germ in ξ at the origin, analytic continuation can be followed along any path issuing from the origin and lying in $\mathbb{C} \setminus \{\xi_{a,b}(z)\}$. We obtain a Riemann surface by considering homotopy classes of such pathes; its *main sheet* corresponds to rectilinear paths and can be identified to the holomorphic star of our germ.

were expanding with respect to h defined by $q = \lambda e^{2\pi i h}$ (and then computing a Borel transform with respect to $h^{1/\tau}$) instead of $t = \frac{q-\lambda}{\lambda}$, and this is responsible for a scaling by a factor $2\pi i$ between the variables ζ and ξ for the Borel transforms. The special singular points $\zeta^+_{k,l}(s)$ can be defined by

$$s + \kappa^{-1}\zeta^+_{k,l}(s) = 2\pi i(k\alpha + l), \qquad k, l \in \mathbb{Z},$$

where $\kappa = \kappa_+(\lambda)$ is the largest number such that $|\frac{N}{D} - \alpha| \geq (\frac{\kappa}{D})^\tau$ for all $\frac{N}{D} > \alpha$ except a finite number of them (recall that $\tau = 2$ in that case). In the resonant case we can set $\kappa = \frac{1}{m_0}$: this is the largest number such that $|\frac{N}{D} - \frac{n_0}{m_0}| \geq \frac{\kappa}{D}$ for all $\frac{N}{D} \neq \alpha$ ($\tau = 1$ in this case and we need not distinguish left and right rational approximations of n_0/m_0). The first line of moving singular points appears to be defined by

$$s + \frac{1}{2\pi i}\kappa^{-1}\xi_{1,b}(z) = -2\pi i\frac{b}{m_0}, \qquad b \in \mathbb{Z},$$

but the group $\{\, -\frac{b}{m_0};\ b \in \mathbb{Z} \,\} = \{\, k\frac{n_0}{m_0} + l;\ k, l \in \mathbb{Z} \,\}$ is discrete, thus the singular points are isolated (hence the resurgence property), whereas $\{\, k\alpha + l;\ k, l \in \mathbb{Z} \,\}$ was dense in $\mathbb{R}$, hence the natural boundary for $\hat{\psi}^+(\zeta, s)$.

REMARK 4.2. If a function $g \in B_{r_1}$ is given, for some $r_1 > 0$, one can deduce results for the corresponding solution $f = f_\delta \odot g$: we know by Theorem 4.1 that $(q - \Lambda_0)f \in \mathcal{G}_1(\Lambda_0, B_{r_2})$ and the function $\hat{\Phi}^g = \mathcal{B} \circ J_{\Lambda_0}\big(-\Lambda_0\mathcal{L}_{m_0} \odot g + (q - \Lambda_0)f\big)$ belongs to $\hat{\mathcal{N}}(B_{r_2})$ for all $r_2 \in]0, r_1[$. In fact, for each $\xi \in \mathbb{C}$, $\hat{\Phi}^g = \hat{\Phi}^\delta \odot g$ and the singularities with respect to ξ of $\hat{\Phi}^g$ depend on the singularities (with respect to z) of g. More precisely, the location of the moving singular points of $\hat{\Phi}^\delta$ shows that $\hat{\Phi}^\delta$ is holomorphic in $\{\, (\xi, z) \mid z \in \mathbb{D},\ |\Im\mathrm{m}\,\xi| < 2\pi \ln\frac{1}{|z|} \,\} = \{\, |z| < \exp(-\frac{|\Im\mathrm{m}\,\xi|}{2\pi}) \,\}$; thus $\hat{\Phi}^g$ is holomorphic in $\{\, |z| < r_1 \exp(-\frac{|\Im\mathrm{m}\,\xi|}{2\pi}) \,\}$, which means that for each $z \in \mathbb{D}_{r_1}$, $\hat{\Phi}^g$ is holomorphic with respect to ξ in a horizontal strip of width $4\pi r_1 \ln\frac{1}{|z|}$. (But $\hat{\Phi}^g$ may have a natural boundary with respect to ξ if this is the case for g with respect to z.)

4.2.2. So far we were dealing with Borel transforms with respect to $t = \frac{q-\Lambda_0}{\Lambda_0}$, but in fact the variable $\eta = \log(q/\Lambda_0)$ is more convenient. Thus we consider the function

$$\eta \mapsto \Psi^g(\eta) = \eta f(\Lambda_0 e^\eta)$$

still for a general solution $f = f_\delta \odot g$: it admits a Gevrey-1 asymptotic expansion

$$\tilde{\Psi}^g(\eta) = \sum_{p \geq 0} \Psi^g_p \eta^p$$

for η tending to zero by the left or by the right, whose constant term is $\Psi^g_0 = \mathcal{L}_{m_0} \odot g$, and we are interested in the Borel transforms

$$\hat{\Psi}^g = \sum_{n \geq 0} \Psi^g_{n+1}\frac{\xi^n}{n!}, \qquad \hat{\Psi}^\delta = \sum_{n \geq 0} \Psi^\delta_{n+1}\frac{\xi^n}{n!}.$$

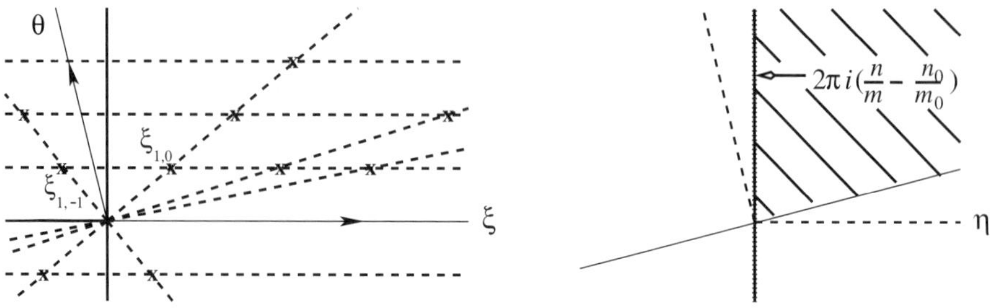

FIGURE 8

THEOREM 4.3 (Borel transform at resonances). *When viewed as a holomorphic function of two variables, $\hat{\Psi}^g$ can be written*

$$\hat{\Psi}^g(\xi, z) = \sum_{k=0}^{m_0-1} \Big(\frac{k}{m_0} - \frac{1}{2}\Big) g(\Lambda_0^k z) - \sum_{k=0}^{m_0-1} \sum_{a\in\mathbb{Z}^*} \frac{e^{2\pi i \frac{ka}{m_0}}}{2\pi i a} \Big[g(\Lambda_0^k z\, e^{\frac{m_0 \xi}{2\pi i a}}) - g(\Lambda_0^k z) \Big]$$

for $|\Im\mathrm{m}\,\xi|$ and $|z|$ small enough. In particular, for each $z \in \mathbb{D}^*$, the function $\xi \mapsto \hat{\Psi}^\delta(\xi, z)$ is meromorphic with simple poles only, located at the points $\xi_{a,b}(z)$, with $C_{a,b}$ as corresponding residues. Moreover, for any $z \in \mathbb{D}^*$ and for any line Δ of $\mathbb{C}$ passing through the origin and avoiding the poles $\xi_{a,b}(z)$, the function $(1 + |\xi|)^{-1}\hat{\Psi}^\delta(\xi, z)$ is bounded for $\xi \in \Delta$.

REMARK 4.3. The knowledge of the residues of $\hat{\Psi}^\delta$ with respect to ξ allows us to compute the "residues" of f_δ with respect to q, i.e. to determine the sequence (a_Λ) such that $f_\delta = \Sigma_{\mathcal{R}}((a_\Lambda))$. In other words the complete asymptotic expansion of f_δ at one resonance contains the information on the leading term in the asymptotics at all other resonances.

Indeed let us fix $\Lambda = e^{2\pi i n/m} \in \mathcal{R}$, with $\frac{n}{m} > \frac{n_0}{m_0}$ for conveniency (and as always $m \in \mathbb{N}^*$, $m \in \mathbb{Z}$, $(n|m) = 1$), and $z \in \mathbb{D}$, $s = \log z$ (the dependence on z of the various functions below will be usually omitted). We will check directly from Theorem 4.3 that $f_\delta(q, z) \sim \frac{a_\Lambda}{q-\Lambda} = \frac{\Lambda}{q-\Lambda}\mathcal{L}_m(z)$ for q tending non-tangentially w.r.t. $\mathbb{S}^1$ to Λ, which is obviously equivalent to

$$f_\delta(e^{2\pi i h}, z) \sim \frac{\mathcal{L}_m(z)}{2\pi i(h - \frac{n}{m})}$$

for h tending non-tangentially w.r.t. $\mathbb{R}$ to $\frac{n}{m}$.

Let us choose a direction θ in $]0, \pi[$ such that $\arg(\xi_{1,0}(z)) < \theta < \arg(\xi_{1,-1}(z))$ (see Figure 8). By Cauchy theorem, we can compare the two Laplace transforms

$$\Psi^\delta(\eta) = \eta f_\delta(\Lambda_0 e^\eta) = \mathcal{L}_{m_0} + \int_0^{+\infty} \hat{\Psi}^\delta(\xi)\, e^{-\xi/\eta}\, d\xi \quad \text{for } \Re\mathrm{e}\,\eta > 0$$

and

$$\Psi_\theta^\delta(\eta) = \int_0^{e^{i\theta}\infty} \hat{\Psi}^\delta(\xi)\, e^{-\xi/\eta}\, d\xi \quad \text{for } \Re\mathrm{e}(\eta\, e^{-i\theta}) > 0.$$

If $\Re e\, \eta > 0$ and $\Re e(\eta\, e^{-i\theta}) > 0$, i.e. η belongs to the intersection of the two half-planes,

$$\eta f_\delta(\Lambda_0 e^\eta) = \mathcal{L}_{m_0} + \Psi_\theta^\delta(\eta) + 2\pi i \sum_{a\geq 1, b\geq 0} C_{a,b}\, e^{-\xi_{a,b}/\eta}.$$

We are interested in η tending to $2\pi i(\frac{n}{m} - \frac{n_0}{m_0})$ from the right. The term $\Psi_\theta^\delta(\eta)$ is regular there and will yield no contribution in the singular behaviour that we want to analyze. On the contrary, for each $a \geq 1$, the sum of the geometric series

$$2\pi i \sum_{b\geq 0} C_{a,b}\, e^{-\xi_{a,b}/\eta} = -\frac{2\pi i}{m_0} \cdot \frac{e^{\frac{2\pi i a s}{m_0 \eta}}}{1 - e^{\frac{2\pi i a}{m_0}(n_0' + \frac{2\pi i}{m_0 \eta})}}$$

defines a function which is meromorphic w.r.t. $1/\eta$. Translating this in the variable $h = \frac{n_0}{m_0} + \frac{\eta}{2\pi i}$ (h tends to n/m with $\Im m\, h < 0$), we obtain

$$(h - \frac{n_0}{m_0}) f_\delta(e^{2\pi i h}) = -\frac{1}{m_0} \sum_{a\geq 1} \frac{e^{\frac{as}{m_0 h - n_0}}}{1 - e^{2\pi i a \frac{n_0' h + m_0'}{m_0 h - n_0}}} + \text{regular function},$$

where we have introduced $m_0' \in \mathbb{Z}$ defined by $m_0 m_0' + n_0 n_0' = 1$.

The image of $\frac{n}{m}$ by the linear fractional map $h \mapsto \frac{n_0' h + m_0'}{m_0 h - n_0}$ is $\frac{N}{M}$, where $N = n_0' n + m_0' m$, $M = m_0 n - n_0 m$ and $(N|M) = 1$. The only terms contributing to the singularity at $h = \frac{n}{m}$ correspond thus to $a = jM$, $j \geq 1$, and an easy computation allows one to conclude that

$$(h - \frac{n_0}{m_0}) f_\delta(e^{2\pi i h}) \sim \frac{1}{2\pi i} \cdot \frac{M}{m_0 m} \cdot \frac{1}{h - \frac{n}{m}} \cdot \sum_{j\geq 1} \frac{e^{jms}}{jm},$$

hence $\lim (h - \frac{n}{m}) f_\delta(e^{2\pi i h}) = \frac{1}{2\pi i} \mathcal{L}_m(z)$.

4.3. Proof of Theorems 4.2 and 4.3

4.3.1. Theorem 4.3 implies Theorem 4.2. This is an exercise of application of the general theory of which Appendix A.5 gives a brief account. We will relate $\hat{\Phi}^\delta(\xi)$ and $\hat{\Psi}^\delta(\xi)$ (from now on we will omit the dependence on the variable z), and first prove that $\hat{\Phi}^\delta$ extends analytically to the universal covering $\mathcal{C}$ of $\mathbb{C} \setminus \{\xi_{a,b}\}$ with at most exponential growth at infinity just because $\hat{\Psi}^\delta(\xi)$ has that property.

In the vicinity of the resonant point Λ_0, we have two local variables t and η:

$$q = \Lambda_0(1+t) = \Lambda_0\, e^\eta,$$

and correspondingly two representations of f_δ as a Laplace transform:

$$t f_\delta = \mathcal{L}_{m_0} + \Lambda_0^{-1} \mathbb{L}_{(\xi\to t)} \hat{\Phi}^\delta, \quad \eta f_\delta = \mathcal{L}_{m_0} + \mathbb{L}_{(\xi\to\eta)} \hat{\Psi}^\delta.$$

We retain that, under the change of variable $t = e^\eta - 1 \Leftrightarrow \eta = \log(1+t)$,

$$\begin{aligned} \mathbb{L}_{(\xi\to t)} \hat{\Phi}^\delta &= \Lambda_0 t L(t) \mathcal{L}_{m_0} + \Lambda_0 (1 + tL(t)) \mathbb{L}_{(\xi\to\eta)} \hat{\Psi}^\delta, \\ L(t) &= \big(\log(1+t)\big)^{-1} - t^{-1} = \frac{1}{2} + O(t). \end{aligned} \tag{4.2}$$

Now we can write $\mathbb{L}_{(\xi\to\eta)}\hat{\Psi}^\delta = \mathbb{L}_{(\xi\to t)}\hat{\chi}$, i.e. we can interpret the change of variable in the Borel plane, by defining $\hat{\chi}(\xi)$ as follows:

$$\hat{\chi}(\xi) = e^{-\xi/2}\hat{\Psi}^\delta(\xi) + \sum_{r\geq 1} \hat{\ell}^{*r} * \frac{\hat{\partial}^r(e^{-\xi/2}\hat{\Psi}^\delta)}{r!},$$

where $\hat{\ell}(\xi)$ is the Borel transform of $\ell(t) = -\frac{1}{2} + L(t)$ and is thus an entire function of exponential type. (This is because $\eta^{-1} = t^{-1} + \frac{1}{2} + \ell(t)$: the translation by $1/2$ is responsible for the multiplication by $e^{-\xi/2}$, and we are then left with *composition-convolution* as described in Appendix A.5. The notation $\hat{\partial}$ simply means multiplication by $-\xi$, the Borel counterpart of differentiation w.r.t. t^{-1}.)

We observe that $\hat{\chi}$ extends analytically to $\mathcal{C}$ with at most exponential growth at infinity, thus this is also the case for

$$\hat{\Phi}^\delta = \Lambda_0\big(\hat{M}\mathcal{L}_{m_0} + \hat{\chi} + \hat{M} * \hat{\chi}\big),$$

where the entire function $\hat{M}$ is simply the Borel transform of $tL(t)$ (thus $\hat{M} = \frac{1}{2} + 1 * \hat{\ell}$).

We must now compute the singularity of $\hat{\Phi}^\delta$ at a point $\omega = \xi_{a,b}$. For that purpose we can use Écalle's formalism of *alien calculus*: in our particular case, the result to be checked is equivalent to the formula

$$\Delta_{(\omega\to t)}\Phi^\delta = 2\pi i\Lambda_0 C_{a,b}(e^{-\omega/2} + L_\omega(t)),$$

whereas the indications of Theorem 4.3 on the poles of $\hat{\Psi}^\delta$ amount to

$$\Delta_{(\omega\to\eta)}\Psi^\delta = 2\pi i C_{a,b}.$$

The operator $\Delta_{(\omega\to t)}$ is the *alien derivation* of index ω relative to the variable t; it is defined so to measure the singular behaviour at ω of the Borel transform w.r.t. t of the function on which it is evaluated. For instance it vanishes on $tL(t)$ since the corresponding Borel transform is entire. The result to be checked is a consequence of the relation (4.2) and of the fact that $\Delta_{(\omega\to t)}$ is a derivation and $e^{-\omega t^{-1}}\Delta_{(\omega\to t)} = e^{-\omega\eta^{-1}}\Delta_{(\omega\to\eta)}$ under the change of variable $\eta^{-1} = t^{-1} + L(t)$. Indeed, when applied to (4.2), these rules imply that

$$e^{-\omega t^{-1}}\Delta_{(\omega\to t)}\Phi^\delta = \Lambda_0(1 + tL(t))e^{-\omega(t^{-1}+L(t))}2\pi i C_{a,b},$$

while precisely $(1 + tL(t))e^{-\omega L(t)} = e^{-\omega/2} + L_\omega(t)$. □

4.3.2. Formal part of the proof of Theorem 4.3. Since $\hat{\Psi}^g = \hat{\Psi}^\delta \odot g$ and $g(\lambda z) = \delta(\lambda z) \odot g(z)$ for all $\lambda \in \mathbb{C}$, it is sufficient to consider the case where $g = \delta$. From now on we will omit the superscript δ. We also replace the variable z by $s = \log z$ (and still keep the same names for some of our functions), so that

$$\Psi(\eta, s) = \eta f_\delta(\Lambda_0 e^\eta, e^s) \sim \tilde{\Psi}(\eta, s) \qquad \text{as } \eta \to 0,$$

and our goal is to study the Borel transform $\hat{\Psi}(\xi, s)$ of that asymptotic series $\tilde{\Psi}(\eta, s)$.

From the cohomological equation satisfied by f_δ, we deduce an equation which admits Ψ as solution (and thus $\tilde{\Psi}$ as formal solution):

$$\Psi(\eta, s + \Omega + \eta) - \Psi(\eta, s) = \eta\,\varphi(s)\,, \tag{4.3}$$

where

$$\Omega = 2\pi i \frac{n_0}{m_0}, \qquad \varphi(s) = \frac{e^s}{1-e^s}.$$

In fact, at this level, one can retain this sole equation and forget everything else.

LEMMA 4.2. *Equation (4.3) admits a unique formal solution*

$$\tilde{\Psi}(\eta, s) = \sum_{p \geq 0} \eta^p \Psi_p(s)$$

with coefficients analytic in $z = e^s$ *and vanishing for* $z = 0$. *This solution is explicitly given by formulas (4.6), (4.4) and (4.5) below; in particular,* $\Psi_0(s) = \mathcal{L}_{m_0}(e^s) = -\frac{1}{m_0}\log(1 - e^{m_0 s})$.

PROOF. Keeping in mind that the solution is required to be $2\pi i$-periodic in s, we introduce the following linear combinations of the Ω-translations of Ψ:

$$\sigma_r(\eta, s) = \sum_{k=0}^{m_0-1} \frac{\Lambda^{-kr}}{m_0} \Psi^{[k]}(\eta, s) \quad \text{for } r = 0, 1, \dots, m_0 - 1$$

$$\Psi^{[k]}(\eta, s) = \Psi(\eta, s + k\Omega) \qquad \text{for } k = 0, 1, \dots, m_0 - 1.$$

The identities

$$\sum_{r=0}^{m_0-1} \frac{\Lambda^{-kr}}{m_0} = \begin{cases} 1 & \text{if } k = 0 \\ 0 & \text{if } k = 1, \dots, m_0 - 1 \end{cases}$$

yield the inverse formulas

$$\Psi^{[k]}(\eta, s) = \sum_{r=0}^{m_0-1} \Lambda^{kr} \sigma_r(\eta, s) \quad \text{for } k = 0, 1, \dots, m_0 - 1.$$

By combining the Ω-translations of Equation (4.3), we obtain the system of equations

$$(*_r) \qquad \Lambda^r \sigma_r(\eta, s + \eta) - \sigma_r(\eta, s) = \eta\, \varphi_{m_0,r}(s)$$

where

$$(4.4) \qquad \varphi_{m_0,r}(s) = \sum_{k=0}^{m_0-1} \frac{\Lambda^{-kr}}{m_0} \varphi(s + k\Omega) = \sum_{\lambda \in \mathcal{R}_{m_0}} \frac{\lambda^{-r}}{m_0} \varphi(s + \log \lambda),$$

for $r = 0, 1, \dots, m_0 - 1$. The left-hand side of Equation $(*_r)$ may be viewed as a "differential operator of infinite order" $(\Lambda^r e^{\eta \partial_s} - \mathrm{Id})$ acting on σ_r. Let us introduce some elementary functions which are analytic at the origin:

$$(4.5) \qquad \Gamma_a(X) = \frac{X}{ae^X - 1} = \sum_{p \geq 0} \gamma_p(a) X^p \quad \text{for } a \in \mathbb{C}^*.$$

Note that $\gamma_0(a) = 0$ if $a \neq 1$, but $\gamma_0(1) = 1$ and in fact

$$\Gamma_1(X) = 1 - \frac{X}{2} - \sum_{l \geq 1} (-1)^l B_l \frac{X^{2l}}{(2l)!}$$

where the coefficients B_l are the Bernoulli numbers.

The functions Γ_a allow us to solve explicitly the system:

$$(*_r) \quad \Leftrightarrow \quad \sigma_r = \partial_s^{-1}\Gamma_{\Lambda^r}(\eta\partial_s)\varphi_{m_0,r} = \gamma_0(\Lambda^r)\partial_s^{-1}\varphi_{m_0,r} + \sum_{p\geq 1}\eta^p\gamma_p(\Lambda^r)\partial_s^{p-1}\varphi_{m_0,r}$$

for $r = 0, 1, \ldots, m_0 - 1$, with the notation ∂_s^{-1} for the unique primitive with respect to s which vanishes when $z = e^s$ vanishes.

Thus, we obtain only one possible formal solution of (4.3):

$$(4.6) \quad \tilde{\Psi} = \partial_s^{-1}\sum_{r=0}^{m_0-1}\Gamma_{\Lambda^r}(\eta\partial_s)\varphi_{m_0,r} = \partial_s^{-1}\varphi_{m_0,0} + \sum_{p\geq 1}\eta^p\partial_s^{p-1}\sum_{r=0}^{m_0-1}\gamma_p(\Lambda^r)\varphi_{m_0,r}$$

Since

$$\partial_s^{-1}\varphi(s) = -\log(1-e^s)\,,$$

we recognize the function $\mathcal{L}_{m_0}$ in the constant term:

$$\begin{aligned}\Psi_0(s) = \partial_s^{-1}\varphi_{m_0,0}(s) &= -\frac{1}{m_0}\log\prod_{k=0}^{m_0-1}(1-e^{s+k\Omega}) \\ &= -\frac{1}{m_0}\log(1-e^{m_0 s})\,.\end{aligned}$$

The formal series $\tilde{\Psi}$ that we just defined is indeed a solution of Equation (4.3): for any $k = 1, \ldots, m_0 - 1$, the formal series

$$\tilde{\Psi}^{[k]}(\eta, s) = \sum_{r=0}^{m_0-1}\Lambda^{kr}\sigma_r(\eta, s)$$

is actually equal to the translation $\tilde{\Psi}(\eta, s+k\Omega)$ of $\tilde{\Psi}$, since for each r the series σ_r is obtained from $\varphi_{m_0,r}$ by applying an operator which commutes with the translations, and

$$\Lambda^{kr}\varphi_{m_0,r}(s) = \varphi_{m_0,r}(s+k\Omega)\,.$$

This remark ends the proof of the lemma. □

REMARK 4.4. The formula that we obtained is reminiscent of the Euler-MacLaurin formula, one of the early sources of divergent asymptotic series. We will analyze it by using the formal Borel transform.[5]

4.3.3. Analytic part of the proof of Theorem 4.3. The above work will now allow us to compute the Borel transform w.r.t. η of $\tilde{\Psi} - \Psi_0$. The starting point is the following decomposition of the functions Γ_{Λ^r} which appear in Formula (4.6):

$$\Gamma_{\Lambda^r}(X) = -\frac{X}{2} + \sum_{\nu\in 2i\pi\mathbb{Z}}^{e}\frac{X}{X + r\Omega - \nu}\,,$$

[5] In [CCD] too Borel transform is used in relation with the Euler-MacLaurin formula, but not with respect to the same variable; our problem pertains rather to *parametric resurgence* according to Écalle's terminology.

where the symbol $\sum^e$ denotes Eisenstein summation [We]: terms corresponding to opposite indices are grouped in order to ensure convergence, i.e.

$$\sum_{l\in\mathbb{Z}}^{e} = \lim_{L\to+\infty} \sum_{l=-L}^{+L} .$$

This decomposition results from the identity

$$\Gamma_{\Lambda^r}(X) = \frac{X}{2}\Big(\coth\frac{X+r\Omega}{2} - 1\Big)$$

and from the classical decomposition

$$\coth X = \sum_{l\in\mathbb{Z}}^{e} \frac{1}{X - il\pi} .$$

It implies that, for $r = 0, 1, \ldots, m_0 - 1$,

$$\sum_{p\geq 0} \gamma_{p+1}(\Lambda^r) X^{p+1} = -\tfrac{1}{2}X - \sum_{\substack{\nu\in 2i\pi\mathbb{Z}\\ \nu\neq 0 \text{ if } r=0}}^{e} \sum_{p\geq 0} (\nu - r\Omega)^{-p-1} X^{p+1},$$

so

$$\sum_{p\geq 0} \gamma_{p+1}(\Lambda^r) \frac{(\xi\partial_s)^p}{p!} = -\tfrac{1}{2}\,\mathrm{Id} - \sum_{\substack{\nu\in 2i\pi\mathbb{Z}\\ \nu\neq 0 \text{ if } r=0}}^{e} (\nu - r\Omega)^{-1} e^{(\nu - r\Omega)^{-1}\xi\partial_s} .$$

According to Formula (4.6) and because of the Taylor formula, the Borel transform of $\tilde{\Psi} - \Psi_0$ can thus be written

$$\hat{\Psi}(\xi, s) = -\sum_{r=0}^{m_0-1}\Big(\tfrac{1}{2}\varphi_{m_0,r}(s) + \sum_{\substack{\nu\in 2i\pi\mathbb{Z}\\ \nu\neq 0 \text{ if } r=0}}^{e} (\nu - r\Omega)^{-1}\varphi_{m_0,r}(s + (\nu - r\Omega)^{-1}\xi)\Big)$$

$$= -\tfrac{1}{2}\varphi(s)$$

$$- \sum_{\substack{l\in\mathbb{Z}, 0\leq r\leq m_0-1\\ (l,r)\neq 0}}^{e} \sum_{k=0}^{m_0-1} \frac{\Lambda^{-kr}}{m_0} (2\pi il - r\Omega)^{-1}\varphi(s + k\Omega + (2\pi il - r\Omega)^{-1}\xi).$$

This is an equality between formal series of powers of ξ, the right-hand side being considered as a formal Taylor expansion (Eisenstein summation ensures that each of its coefficients is well defined). But we can now identify the right-hand side with a series of meromorphic functions, which is easily seen to be convergent since φ and φ' are bounded in any domain of $\mathbb{C}$ obtained by removing small disks around their poles. So we can conclude that $\hat{\Psi}$ converges at the origin and extends to a meromorphic function. The convergence can be made more obvious and the expression of $\hat{\Psi}$ more convenient; we will give these details now.

The value of $\hat{\Psi}$ at $\xi = 0$ is already known from (4.6):

$$\Psi_1 = \sum_{r=0}^{m_0-1} \gamma_1(\Lambda^r)\varphi_{m_0,r} = -\tfrac{1}{2}\varphi_{m_0,0} - \sum_{r=1}^{m_0-1} \frac{1}{1-\Lambda^r}\varphi_{m_0,r},$$

so we have now two expressions for it:
(4.7)

$$\hat{\Psi}(0,s) = \Psi_1(s) = -\frac{1}{m_0}\sum_{k=0}^{m_0-1}\Big(\frac{1}{2} + \sum_{r=1}^{m_0-1}\frac{\Lambda^{-kr}}{1-\Lambda^r}\Big)\varphi(s+k\Omega)$$

$$= -\tfrac{1}{2}\varphi(s) - \sideset{}{^e}\sum_{\substack{l\in\mathbb{Z},0\le r\le m_0-1\\(l,r)\ne 0}}\sum_{k=0}^{m_0-1}\frac{\Lambda^{-kr}}{m_0}(2\pi il - r\Omega)^{-1}\varphi(s+k\Omega).$$

Substracting it from $\hat{\Psi}$, we can write a uniformly convergent sum

$$\hat{\Psi}(\xi,s) - \Psi_1(s) =$$

$$\sum_{\substack{0\le r\le m_0-1\\ l\in\mathbb{Z};\,(l,r)\ne 0}}\sum_{k=0}^{m_0-1}\frac{\Lambda^{-kr}}{m_0}(2\pi il - r\Omega)^{-1}\left(\varphi(s+k\Omega+(2\pi il-r\Omega)^{-1}\xi) - \varphi(s+k\Omega)\right)$$

without using Eisenstein summation.

We have $\Omega = 2\pi in_0/m_0$ with $m_0m_0' + n_0n_0' = 1$ for some integers m_0', n_0'. The application

$$\begin{cases} \mathbb{Z}\times\{0,\dots,m_0-1\} \longrightarrow \quad \mathbb{Z} \\ \qquad (l,r) \qquad \longmapsto a = lm_0 - rn_0 \end{cases}$$

is a bijection, the inverse of which is given by

$$l = am_0' - cn_0\,, \quad r = -an_0' - cm_0$$

where c is the integer part of $-an_0'/m_0$. Thus, we can use it as a change of indices: $2\pi il - r\Omega = 2\pi ia/m_0$, and $\Lambda^{-kr} = e^{2\pi ika/m_0}$ because $rn_0 \equiv -a (\mathrm{mod}\ m_0)$, so we end up with the formula

$$(4.8)\quad \hat{\Psi}(\xi,s) = \Psi_1(s) - \sum_{a\in\mathbb{Z}^*}\sum_{k=0}^{m_0-1}\frac{1}{2\pi ia}e^{2\pi ika/m_0}\Big(\varphi(s+k\Omega+\frac{m_0\xi}{2\pi ia}) - \varphi(s+k\Omega)\Big).$$

Here is an argument for proving that this series converges uniformly and defines a function which is meromorphic with respect to ξ for $\Re e\, s < 0$: it is sufficient to check, for any positive constant ρ, the uniform convergence in the set

$$E_\rho = \{\,(\xi,s)\in\mathbb{C}\times\mathbb{C} \mid\ \Re e\, s \le -\rho \text{ and } \forall a\in\mathbb{Z}^*,\ \forall b\in\mathbb{Z},\ |\xi - \xi_{a,b}(s)| \ge |a|\rho\,\}$$

(working in E_ρ means removing a small disk around each singularity in the ξ-plane). Let us fix ρ and define the set

$$\mathcal{D}_\rho = \{\, s\in\mathbb{C} \mid\ \forall l\in\mathbb{Z},\ |s - 2\pi il| \ge m_0\rho/2\pi\,\}$$

so to have the following relation between E_ρ and $\mathcal{D}_\rho$:

$$(\xi,s)\in E_\rho \quad \Leftrightarrow \quad \begin{cases} \Re e\, s \le -\rho, \\ s + k\Omega + \frac{m_0\xi}{2\pi ia} \in \mathcal{D}_\rho \qquad \text{for } 0\le k\le m_0-1 \text{ and } a\in\mathbb{Z}^*; \end{cases}$$

note that $\Re e\, s \le -\rho$ implies that the points $s+k\Omega$ belong to $\mathcal{D}_\rho$ too. The function φ is $2\pi i$-periodic and its derivative is bounded in $\mathcal{D}_\rho$; there exists $c_\rho > 0$ such that any two points s and s' in $\mathcal{D}_\rho$ can be joined inside $\mathcal{D}_\rho$ by a path of length less than $c_\rho|s-s'|$ followed by an integer number of $2\pi i$-translations, hence

$$\forall s, s' \in \mathcal{D}_\rho, \quad |\varphi(s') - \varphi(s)| \leq M_\rho|s-s'| \quad \text{with} \quad M_\rho = c_\rho \sup\{\, |\varphi'(s)|, \ s \in \mathcal{D}_\rho \,\}.$$

This implies the uniform convergence of our series, with an explicit bound

$$\forall (\xi, s) \in E_\rho, \quad |\hat{\Psi}(\xi, s)| \leq |\Psi_1(s)| + |\xi| \sum_{a\in\mathbb{Z}^*} \frac{{m_0}^2 M_\rho}{4\pi^2|a|^2}$$

which shows the slow growth of $\hat{\Psi}$ with respect to ξ. Note that the function Ψ_1 is bounded in $\mathcal{D}_\rho$ (since φ is bounded in $\mathcal{D}_\rho$).

The function φ is meromorphic with only simple poles, located at the points $2\pi i l$ for $l \in \mathbb{Z}$; the corresponding residue is -1. Thus, for fixed s, the function $\varphi(s + k\Omega + \frac{m_0\xi}{2\pi i a})$ is meromorphic with respect to ξ, with only simple poles located at the points

$$\frac{2\pi i a}{m_0}\left(-s + \frac{2\pi i}{m_0}(l m_0 - k n_0)\right) = \xi_{a,b}(s)$$

with $b = -l m_0 + k n_0$; the corresponding residue is $2\pi i a/m_0$ and $k \equiv b n_0'(\text{mod } m_0)$, hence the value of the residue of $\hat{\Psi}$ at $\xi_{a,b}(s)$.

We let the reader check that

$$\Psi_1(s) = \sum_{k=0}^{m_0-1} \Big(\frac{k}{m_0} - \frac{1}{2}\Big)\, \varphi(s + k\Omega)\,. \tag{4.9}$$

from the identity (4.7). □

REMARK 4.5. Using again a decomposition formula, but this time for φ:

$$\varphi(s) = 1 + \frac{1}{e^s - 1} = \frac{1}{2} + \sum_{\nu\in 2i\pi\mathbb{Z}}^{e} \frac{1}{s-\nu},$$

one finds the formula

$$\hat{\Psi}(\xi, s) = \Psi_1(s) + \sum_{a\in\mathbb{Z}^*,\, b\in\mathbb{Z}} C_{a,b}\left(\frac{1}{\xi - \xi_{a,b}(s)} + \frac{1}{\xi_{a,b}(s)}\right)$$

with uniform convergence in any compact subset of E_ρ.

REMARK 4.6. One can write a different proof of Theorem 4.3 by starting from the decomposition of f_δ as a sum of simple poles. We prefered to use a method which relies only on Equation (4.3) because it can be adapted in some nonlinear problems (see Section 5.3).

4.4. A property of quasianalyticity at constant-type points

4.4.1. Let us fix $\lambda = e^{2\pi i \alpha} \in \mathbb{S}^1$ with $\alpha \in [0, 1[$ and a Banach space B. We now introduce spaces of functions which admit Gevrey asymptotics inside cardioids with cusp at λ.

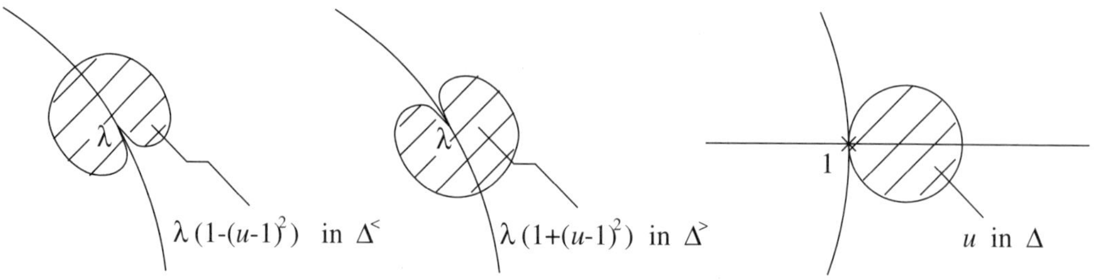

FIGURE 9

DEFINITION 4.2. For any $\tau > 0$, we define $\mathcal{G}_\tau^<(\lambda, B)$ to be the space of all B-valued functions f such that $u \mapsto f(\lambda(1-(u-1)^2))$ defines a function of $\mathcal{G}_\tau^+(1, B)$. Analogously, we define $\mathcal{G}_\tau^>(\lambda, B)$ to be the space of all B-valued functions f such that $u \mapsto f(\lambda(1+(u-1)^2))$ defines a function of $\mathcal{G}_\tau^+(1, B)$.

Equivalently, $\mathcal{G}_\tau^<(\lambda, B)$ or $\mathcal{G}_\tau^>(\lambda, B)$ is the set of the functions f which are analytic in some open set whose boundary is a cardioid $\Delta^<$ or $\Delta^>$ with its cusp at λ and its axis tangent to $\mathbb{S}^1$ at λ, oriented according to Figure 9 (such a cardioid $\Delta^{\gtrless}$ is nothing but the image of some disk Δ by $u \mapsto \lambda(1 \pm (u-1)^2)$), and for which there exist a formal series $\sum_{n \geq 0} a_n Q^n \in B[[Q]]$ and positive numbers c_0, c_1 such that

$$\forall N \geq 0,\ \forall q \in \Delta^{\gtrless}, \quad \|f(q) - \sum_{0 \leq n \leq N-1} a_n (q-\lambda)^n\| \leq c_0\, c_1^N\, \Gamma(1+2\tau N)\, |q-\lambda|^N.$$

Thus such a function admits Gevrey-2τ asymptotics inside the cardioid. In particular, for $\tau = 1$, we observe that $\mathcal{G}_1^<(\lambda, B)$ and $\mathcal{G}_1^>(\lambda, B)$ are quasianalytic spaces whose members admit Gevrey-2 asymptotics at λ.

DEFINITION 4.3. We define two mappings $\Sigma_\lambda^<,\ \Sigma_\lambda^> : \ell^1(\mathcal{R}, B) \to \mathcal{O}(\mathbb{D} \cup \mathbb{E}, B)$ by the formulas

$$\Sigma_\lambda^{\gtrless}(a)(q) = \sum_{\Lambda \in \mathcal{R} \cap \mathbb{S}_\lambda^{\gtrless}} \frac{a_\Lambda}{q - \Lambda} \qquad \text{if} \quad a = (a_\Lambda)_{\Lambda \in \mathcal{R}} \in \ell^1(\mathcal{R}, B),$$

where $\mathbb{S}_\lambda^< = \{\, e^{2\pi i x},\ x \in]\alpha - 1/2, \alpha[\,\}$ and $\mathbb{S}_\lambda^> = \{\, e^{2\pi i x},\ x \in]\alpha, \alpha + 1/2[\,\}$.

This way, we obtain a decomposition of any Borel-Wolff-Denjoy series with poles in $\mathcal{R}$: if $\lambda \notin \mathcal{R}$, $\Sigma_{\mathcal{R}} = \Sigma_\lambda^< + \Sigma_\lambda^>$ (if $\lambda \in \mathcal{R}$, one should add the contributions of λ and $-\lambda$). This is quite reminiscent of the decomposition of the fundamental solution at the beginning of the proof of Theorem 3.5, except that the starting point there was Lemma 3.4 which decomposes the function according to its poles with respect to $h = \frac{1}{2\pi i} \log \frac{q}{\lambda}$ rather than with respect to q.

LEMMA 4.3. *Let $r \in]0,1[$. If $\lambda \in \underline{\mathrm{DC}}_\tau$ with $\tau = 1$ or $\tau \geq 2$, the inclusions $\Sigma_\lambda^<(\mathcal{S}(r,B)) \subset \mathcal{G}_{\tau/2}^<(\lambda, B)$ and $\Sigma_\lambda^>(\mathcal{S}(r,B)) \subset \mathcal{G}_{\tau/2}^>(\lambda, B)$ hold.*

PROOF. Follow the lines of the proof of Theorem 4.1, in particular adapt Lemma 4.1 and choose for $\mathcal{K}$ a compact set bounded by a cardioid ($\beta = 3/2$). □

For our purpose the previous lemma will not be of any particular interest for $\tau = 1$, i.e. for resonant points, whereas for $\tau = 2$ it has the advantage of letting appear

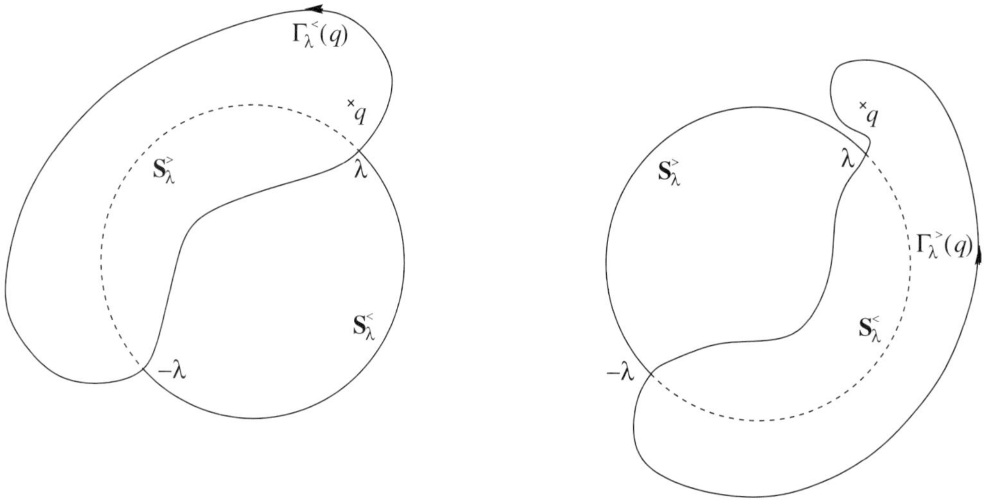

FIGURE 10

the quasianalytic spaces $\mathcal{G}_1^{\gtrless}(\lambda, B)$ in connection with constant-type points. But of course, for a given Borel-Wolff-Denjoy series $f = \Sigma_{\mathcal{R}}(a)$, instead of dealing with f itself that result only tells that two series $\Sigma_\lambda^<(a)$ and $\Sigma_\lambda^>(a)$, whose sum is f, belong to $\mathcal{G}_1^<(\lambda, B)$ or $\mathcal{G}_1^>(\lambda, B)$, and adding functions belonging to different quasianalytic classes is known to be a delicate matter (cf. Mandelbrojt's theorem quoted in [Th] or [E3], but also [P2]). In fact, in our situation, the relevant question is to know whether we can recover the series $\Sigma_\lambda^<(a)$ and $\Sigma_\lambda^>(a)$ directly from f.

4.4.2. A first answer is provided by the

LEMMA 4.4. *Assume $\lambda \in \underline{\mathrm{DC}}_\tau$ with $\tau \geq 2$. Let $r \in]0,1[$, $a \in \mathcal{S}(r, B)$ and $q \in \mathbb{D} \cup \mathbb{E}$. One can write*

$$\Sigma_\lambda^<(a)(q) = \frac{1}{2\pi i} \int_{\Gamma_\lambda^<(q)} \frac{\Sigma_{\mathcal{R}}(a)(q_1)}{q_1 - q} \, dq_1, \quad \Sigma_\lambda^>(a)(q) = \frac{1}{2\pi i} \int_{\Gamma_\lambda^>(q)} \frac{\Sigma_{\mathcal{R}}(a)(q_1)}{q_1 - q} \, dq_1,$$

if $\Gamma_\lambda^<(q)$—resp. $\Gamma_\lambda^>(q)$)— is a simple loop with anticlockwise orientation, intersecting $\mathbb{S}^1$ at λ and $-\lambda$ only, transversally, and enclosing the point q and the set $\mathbb{S}_\lambda^>$— resp. the set $\mathbb{S}_\lambda^<$—(see Figure 10).

The proof of Lemma 4.4 is left to the reader.

But the formulas above are "global" with respect to q, in the sense that $(\Sigma_\lambda^<(a)(q))(z)$ and $(\Sigma_\lambda^<(a)(q))(z)$ depend there on the numbers $(\Sigma_{\mathcal{R}}(a)(q_1))(z)$. It would be more interesting to have formulas which are local in q and global in z. This turns out to be possible *when restricting to solutions of the cohomological equation.*

LEMMA 4.5. *Let $\lambda \in \mathbb{S}^1 \setminus \mathcal{R}$ and define the coefficients*

$$\delta_{n,\ell}^<(\lambda) = \frac{1}{n} \sum_{\Lambda \in \mathcal{R}_n \cap \mathbb{S}_\lambda^<} \Lambda^{-\ell}, \quad \delta_{n,\ell}^>(\lambda) = \frac{1}{n} \sum_{\Lambda \in \mathcal{R}_n \cap \mathbb{S}_\lambda^>} \Lambda^{-\ell}, \qquad n \geq 1,\ n-1 \geq \ell \geq 0$$

(recall that $\mathcal{R}_n = \{\Lambda \in \mathbb{C} \mid \Lambda^n = 1\}$*) and let* $r \in]0,1[$*. For each* $q \in \mathbb{D}_{1/r}$*, the formulas*

$$\delta_\lambda^{\gtrless}(q) \,:\, z \mapsto \sum_{\ell \geq 0, n \geq \ell+1} \delta_{n,\ell}^{\gtrless}(\lambda) q^\ell \, z^n$$

define two members $\delta_\lambda^{<}(q)$ *and* $\delta_\lambda^{>}(q)$ *of* $B_r = zH^\infty(\mathbb{D}_r)$*. The functions* $\delta_\lambda^{<}$ *and* $\delta_\lambda^{>}$ *are* B_r*-valued holomorphic functions in* $\mathbb{D}_{1/r}$ *which satisfy*

$$\forall q \in \mathbb{D}_{1/r}, \quad \delta_\lambda^{<}(q) + \delta_\lambda^{>}(q) = \delta.$$

LEMMA 4.6. *Let us suppose that* $\lambda \in \underline{\mathrm{DC}}_\tau$ *with* $\tau \geq 2$, $0 < r_2 < r_1$ *and* $r \in [r_2/r_1, 1[$*. Let* $g \in B_{r_1}$ *and consider the corresponding solution* $f = f_g$*, written as* $f = \Sigma_{\mathcal{R}}(a)$ *where* $a \in \mathcal{S}(r, B_{r_2})$*: for all* $q \in \mathbb{D}_{1/r} \setminus \mathbb{S}^1$,

$$\Sigma_\lambda^{<}(a)(q) = \delta_\lambda^{<}(q) \odot \Sigma_{\mathcal{R}}(a)(q), \quad \Sigma_\lambda^{<}(a)(q) = \delta_\lambda^{<}(q) \odot \Sigma_{\mathcal{R}}(a)(q).$$

PROOF OF LEMMA 4.5. Let $n \geq 1$ and $0 \leq \ell \leq n-1$. We have obviously $|\delta_{n,\ell}^{\gtrless}(\lambda)| \leq 1$ and

$$\delta_{n,\ell}^{<}(\lambda) + \delta_{n,\ell}^{>}(\lambda) = \begin{cases} 1 & \text{if } \ell = 0, \\ 0 & \text{if } 1 \leq \ell \leq n-1. \end{cases}$$

The Taylor series $\sum_{\ell \geq 0, n \geq \ell+1} \delta_{n,\ell}^{\gtrless}(\lambda) q^\ell \, z^n$ can be written $zE^{\gtrless}(qz, z)$ with a series

$$E^{\gtrless}(x, z) = \sum_{\ell \geq 0, r \geq 0} \delta_{\ell+1+r,\ell}^{\gtrless}(\lambda) x^\ell \, z^n$$

which is convergent for $(x, z) \in \mathbb{D} \times \mathbb{D}$. Thus we get functions which are holomorphic for $(q, z) \in \mathbb{D}_{1/r} \times \mathbb{D}_r$, and for each $q \in \mathbb{D}_{1/r}$ we get functions $\delta_\lambda^{<}(q)$ and $\delta_\lambda^{>}(q)$ which belong to B_r and whose sum is constant and equal to δ. □

PROOF OF LEMMA 4.6. It is sufficient to the consider the case of the fundamental solution, i.e. to prove those identities for $a = \delta$. In that case, $\Sigma_{\mathcal{R}}(a) = f_\delta$ and

$$\Sigma_\lambda^{\gtrless}(a) = \sum_{\Lambda \in \mathcal{R} \cap \mathbb{S}_\lambda^{\gtrless}} \frac{\Lambda}{q - \Lambda} \mathcal{L}_{m(\Lambda)}(z) = \sum_{n \geq 1} A_{n,\lambda}^{\gtrless}(q) z^n,$$

with Taylor coefficients which can written

$$A_{n,\lambda}^{\gtrless}(q) = \frac{1}{n} \sum_{\Lambda \in \mathcal{R}_n \cap \mathbb{S}_\lambda^{\gtrless}} \frac{\Lambda}{q - \Lambda}$$

(because $\mathcal{L}_m(z) = \displaystyle\sum_{n \geq m \text{ s.t. } m|n} \frac{z^n}{n}$). The identities to be proved amount to

$$\forall n \geq 1, \qquad A_{n,\lambda}^{\gtrless}(q) = \frac{1}{q^n - 1} \sum_{\ell=0}^{n-1} \delta_{n,\ell}^{\gtrless}(\lambda) q^\ell,$$

which is easy to check. □

DEFINITION 4.4. For $\tau > 0$ and $r > 0$, we define $\mathcal{G}_\tau^{\odot}(\lambda, B_r)$ to be the subspace of $\mathcal{G}_{2\tau}(\lambda, B_r)$ consisting of all the functions f such that $f \odot \delta_\lambda^<$ extends to a function of $\mathcal{G}_\tau^<(\lambda, B_r)$ and $f \odot \delta_\lambda^>$ extends to a function of $\mathcal{G}_\tau^>(\lambda, B_r)$.

Putting things together we obtain

THEOREM 4.4 (Quasianalyticity at constant-type points). *Let $\lambda \in \underline{\mathrm{DC}}_\tau$ with $\tau \geq 2$. For each $r \in]0, 1[$, the fundamental solution f_δ belongs to $\mathcal{G}_{\tau/2}^{\odot}(\lambda, B_r)$, which is quasianalytic at λ if $\tau = 2$. Thus, if $0 < r_2 < r_1$ and $g \in B_{r_1}$, the corresponding solution f_g belongs to the space $\mathcal{G}_{\tau/2}^{\odot}(\lambda, B_{r_2})$, which is quasianalytic at λ if $\tau = 2$.*

This means in particular that a solution f can be recovered from its asymptotic expansion $\tilde{f}$ at a constant-type point λ by computing and "resumming" independently the series $\tilde{f} \odot \delta_\lambda^<$ and $\tilde{f} \odot \delta_\lambda^>$.

CHAPTER 5

Conclusions and Applications

In this final chapter we first describe an unexpected connection of our work with a conjecture by Gammel. Then we apply the results of Section 3.2 to the problem of linearization of analytic diffeomorphisms of the circle and we briefly sketch how the results of Section 4.2 can be generalized to a nonlinear small divisor problem.

5.1. Gammel's series

In a paper [Gam] published in 1974 Gammel studied the convergence of Padé approximants to quasianalytic functions beyond natural boundaries (see also [GN]). In particular he considered the Borel-Wolff-Denjoy series

$$G(q) = \sum_{m=2}^{\infty} \sum_{\Lambda \in \mathcal{R}_m^*} \frac{e^{-m}}{q - \Lambda}. \tag{5.1}$$

As we have seen in our discussion in Section 2.2 this defines two complex-valued holomorphic functions, one in $\mathbb{D}$ and the other in $\mathbb{E}$, which have the unit circle as a natural boundary of analyticity. Gammel asked whether the function defined in $\mathbb{D}$ could be continued to the one defined in $\mathbb{E}$ through the natural boundary, as his numerical results suggested.[6]

Here we want to show how our results give an affirmative answer to this question, but we leave untouched the quetion of the connection between convergence of Padé approximants and quasianalyticity.[7]

THEOREM 5.1. *There exist $r > 1$ and $g \in B_r$ such that the function $\chi(q, z) = q^{-1}(f_g(q, z) - f_g(0, 1))$ satisfies $\chi(q, 1) = G(q)$ for all $q \in \mathbb{D} \cup \mathbb{E}$. As a consequence,*
i) for all $\Lambda_0 \in \mathcal{R}$, Gammel's series G belongs to the space $(q - \Lambda_0)^{-1}\mathcal{G}_1(\Lambda_0, \mathbb{C})$, which is quasianalytic at Λ_0;
ii) for all $\lambda \in \underline{\mathrm{DC}}_2$ and $r' \in]1, r[$, the function χ belongs to the space $\mathcal{G}_1^{\odot}(\lambda, B_{r'})$, which is quasianalytic at λ.

[6] More precisely, Gammel asked whether the series (5.1) belongs to some quasianalytic space of Borel-Wolff-Denjoy series, and he showed numerically that the Padé approximants $[N/N+1]$ of G at $q = 0$ compute the value of G at $q = 2$ within numerical accuracy. Since the Padé approximants depend only on the Taylor series of G at $q = 0$, this suggested that one could continue quasianalytically G byond its natural boundary $\mathbb{S}^1$.

[7] Gammel's numerical results showing convergence of Padé approximants of G beyond its circle of convergence could probably be justified by adapting [GN] (which deals with the classical quasianalytic class of Borel-Wolff-Denjoy series of the form $\sum_{\nu=1}^{\infty} \frac{A_\nu}{1 - q\alpha_\nu}$, with α_ν dense on the unit circle but $|A_\nu| \le Ce^{-\nu^{1+\varepsilon}}$ for some $\varepsilon > 0$, which is not true for $G(q)$ which has $|A_\nu| \approx \exp(-\sqrt{\nu})$).

All the results on the Whitney smoothness and monogenic dependence with respect to q proved in the previous sections apply to the function χ, thus to Gammel's series G.

As for quasianalyticity, Part *i)* shows that the function G in $\mathbb{E}$ can be recovered from the knowledge of $G_{|\mathbb{D}}$: one can choose any resonance Λ_0 and use Borel-Laplace summation of the asymptotic expansion at Λ_0. Part *ii)* yields another possibility, using the asymptotic expansion at any constant-type point λ, but for $\chi(q,z)$ rather than for G itself: the dependence on z is essential for that kind of quasianalyticity.

PROOF. Let $A_1 = 0$, $A_m = e^{-m}$ for $m \geq 2$. Denoting by φ Euler's totient function, $\varphi(m) = \operatorname{card} \mathcal{R}_m^*$, we have

$$\sum_{\Lambda\in\mathcal{R}_m^*} \frac{1}{q-\Lambda} = q^{-1}\Big(\varphi(m) + \sum_{\Lambda\in\mathcal{R}_m^*} \frac{\Lambda}{q-\Lambda}\Big),$$

thus

$$G(q) = q^{-1}(F(q) - F(0)), \quad \text{with } F(q) = \sum_{m\geq 1}\sum_{\Lambda\in\mathcal{R}_m^*} \frac{\Lambda A_m}{q-\Lambda}.$$

In view of Proposition A2.1, we only need to find $g(z) = \sum_{n\geq 1} g_n z^n$ such that

$$A_m = (g \odot \mathcal{L}_m)_{|z=1} = \sum_{j\geq 1} \frac{g_{mj}}{mj}, \qquad m \geq 1. \tag{5.2}$$

Since $\sum m|A_m| < \infty$, we can use Möbius inversion formula ([HW], Theorem 270, p. 237): we set

$$g_n = n \sum_{j\geq 1} \mu(j) A_{nj}, \quad n \geq 1,$$

where the Möbius function $\mu(j)$ is defined by 1 if $j = 1$, $(-1)^r$ if j is the product of r distinct primes, and 0 if j has a squared factor. This yields a solution of (5.2), because of the relation $\sum_{d|n} \mu(d) = 1$ if $n = 1$ and 0 if $n \geq 2$. We observe that the radius of convergence of $g(z)$ is > 1. We have $F(q) = f_g(q,1)$, thus we set $\chi = q^{-1}(f_g(q,z) - F(0))$ and we can apply Theorems 4.1 and 4.4. □

In the previous example, one can check moreover that $g(z)$ has a radius of convergence equal to e and that it defines a meromorphic function:

$$g(z) = -e^{-1} z + z \sum_{j\geq 1} \mu(j) \frac{e^{-j}}{(1 - z\,e^{-j})^2}.$$

The constant $\chi(0,1)$ involved in the description of $G(q)$ is

$$\sum_{m=2}^{\infty} e^{-m}\varphi(m) = 0.311413131378555402046127705506\ldots$$

As is easily seen from the above proof, the statement of Theorem 5.1 holds for any series

$$G(q) = \sum_{m=2}^{\infty} \sum_{\Lambda\in\mathcal{R}_m^*} \frac{A_m}{q-\Lambda}$$

with $\limsup |A_m|^{1/m} < 1$. But Gammel studies also in his paper the example corresponding to $A_m = e^{-\sqrt{m}}$, for which quasianalyticity seems to fail as well as the convergence of Padé approximants. Indeed, in that case, or more generally if $\sum m|A_m| < \infty$ but $\limsup |A_m|^{1/m} = 1$, our arguments do not apply any longer: there is a series $g(z)$ such that $G(q) = q^{-1}(f_g(q,1) - f_g(0,1))$, but it has a radius of convergence equal to 1, which prevents us to take $r > 1$ and thus to conclude anything for those series.

5.2. An application to the problem of linearization of analytic diffeomorphisms of the circle

As already mentioned in the introduction, the problem of the local conjugacy of analytic diffeomorphisms of the circle leads to the linearized equation (1.3). Here we show how one can use the results of Section 3.2 on the existence at Diophantine points of Gevrey asymptotic expansions of monogenic functions in order to make a recent result of E. Risler [Ris] more precise.

Let $\Delta > 0$, $\varepsilon > 0$, $\alpha \in \mathbb{C}$, $\mu \in \mathbb{C}$. Following [Ris] we define:

$$\begin{aligned}
\mathbb{B}_\Delta &= \{\, z \in \mathbb{C} \mid |\Im\mathrm{m}\, z| < \Delta \,\}, \\
\mathbb{B}_\Delta(\alpha) &= \{\, z \in \mathbb{C} \mid -\Delta < \Im\mathrm{m}\, z < \Delta + \Im\mathrm{m}\, \alpha \text{ if } \Im\mathrm{m}\, \alpha \geq 0, \\
&\qquad\qquad - \Delta + \Im\mathrm{m}\, \alpha < \Im\mathrm{m}\, z < \Delta \text{ if } \Im\mathrm{m}\, \alpha \leq 0 \,\}, \\
\mathcal{D}(\Delta) &= \{\, G : \mathbb{B}_\Delta \to \mathbb{C} \text{ analytic and commuting with integer translations} \,\}, \\
\mathcal{D}(\Delta, \alpha) &= \{\, G : \mathbb{B}_\Delta(\alpha) \to \mathbb{C} \text{ analytic and commuting with integer translations} \,\}, \\
\mathcal{D}_\mu(\Delta) &= \{\, G \in \mathcal{D}(\Delta) \mid \int_0^1 (G(z) - z)dz = \mu \,\}, \\
\mathcal{D}_\mu^\varepsilon(\Delta) &= \{\, G \in \mathcal{D}_\mu(\Delta) \mid \sup_{z \in \mathbb{B}_\Delta} |G(z) - z - \mu| < \varepsilon \,\}, \\
\mathcal{D}^\varepsilon(\Delta) &= \bigcup_{\mu \in \mathbb{C}} \mathcal{D}_\mu^\varepsilon(\Delta), \\
\mathcal{D}_\mu(\Delta, \alpha) &= \{\, G \in \mathcal{D}(\Delta, \alpha) \mid \int_0^1 (G(z) - z)dz = \mu \,\}, \\
\mathcal{D}_\mu^\varepsilon(\Delta, \alpha) &= \{\, G \in \mathcal{D}_\mu(\Delta, \alpha) \mid \sup_{z \in \mathbb{B}_\Delta(\alpha)} |G(z) - z - \mu| < \varepsilon \,\}, \\
\mathcal{D}^\varepsilon(\Delta, \alpha) &= \bigcup_{\mu \in \mathbb{C}} \mathcal{D}_\mu^\varepsilon(\Delta, \alpha).
\end{aligned}$$

We will denote with $\mathcal{D}_\mu^{\varepsilon, inj}(\Delta, \alpha)$ the set of maps in $\mathcal{D}_\mu^\varepsilon(\Delta, \alpha)$ which are injective on $\mathbb{B}_\Delta(\alpha)$.

Let $\gamma > 0$, $\kappa > 0$, $d > 0$ and $\beta > 0$. We consider the approximation function

$$\psi(m) = \gamma \exp\Big(-\frac{m}{(\log m)^{1+\beta}} \Big), \tag{5.3}$$

and the associated domain $C_{\psi,\kappa,d}$ as in Definition 2.4. We retain from Theorem 4, p. 12 of [Ris], the following slightly weaker result:

THEOREM 5.2 (Local conjugacy of analytic diffeomorphisms of the circle with real or complex rotation numbers). *For all $\Delta > \delta > 0$ there exist $\varepsilon > 0$ and a continuous map*

$$(\alpha, F) \in C_{\psi,\kappa,d} \times \mathcal{D}^\varepsilon(\Delta) \mapsto (\ell(\alpha, F), h_{\alpha,F}) \in \mathbb{C} \times \mathcal{D}_\mu^{\delta, inj}(\Delta - \delta, \alpha) \tag{5.4}$$

such that for all $(\alpha, F) \in C_{\psi,\kappa,d} \times \mathcal{D}^\varepsilon(\Delta)$ *and for all* $z \in \mathbb{B}_{\Delta-\delta}$ *one has*

$$\ell(\alpha, F) + F(h_{\alpha,F}(z)) = h_{\alpha,F}(z+\alpha). \tag{5.5}$$

Moreover the map (5.4) *is analytic on* $\mathrm{int}(C_{\psi,\kappa,d}) \times \mathcal{D}^\varepsilon(\Delta)$ *and, for all* $F \in \mathcal{D}^\varepsilon(\Delta)$, *the function*

$$\ell_F \,:\, \alpha \in C_{\psi,\kappa,d} \;\mapsto\; \ell(\alpha, F) \in \mathbb{C} \tag{5.6}$$

is $\mathcal{C}^\infty$*-holomorphic.*

Theorem 5.2 is indeed a generalization of Yoccoz's theorem [Y1,Y2,Y3] on the linearization of analytic diffeomorphisms of the circle close to rotations (inasmuch as rotation numbers are allowed to be complex) and of Herman's theorem ([He]; see also Arnold [Ar]) since the required arithmetical condition is weaker (in [He] the real rotation numbers are assumed to be Diophantine of exponent $\tau \in [0,1]$). The statement in [Ris] is slightly more general than Theorem 5.2 since, instead of using an approximation function, the real rotation numbers belong to any fixed relatively compact subset of the set of Brjuno numbers (w.r.t. a topology, finer than the topology induced by the usual one of $\mathbb{R}$, induced by the embedding of the Brjuno numbers into the space ℓ^1 of summable sequences: see [Ris, pp. 6–9] for details).

The choice of the two positive constants γ and β in the definition of the approximation function (5.3) is arbitrary. Let ψ_j denote the approximation function obtained choosing $\gamma = \gamma_j$, $\beta = \beta_j$ where $(\gamma_j)_{j\in\mathbb{N}}$ and $(\beta_j)_{j\in\mathbb{N}}$ are two positive decreasing sequences which tend to 0. From the previous theorem it follows that

$$\ell_F \in \mathcal{M}((K_j)_{j\in\mathbb{N}}, \mathbb{C}), \qquad K_j = C_{\psi_j,\kappa,d}.$$

We define the Gevrey classes $\tilde{\mathcal{G}}_\tau(y,\mathbb{C})$ for $\tau > 0$ and $y \in \mathbb{R}$ simply by substituting the unit circle $\mathbb{S}^1$ with the real line in Definitions 3.1 and 3.2.

THEOREM 5.3. *Let* $y \in \mathrm{DC}_\tau$. *The function* ℓ_F *belongs to the space* $\tilde{\mathcal{G}}_{\tau'}(y,\mathbb{C})$ *for all* $\tau' > \tau$.

PROOF. This is a minor adaptation of Theorem 3.3. Following Section 3.2 very closely, it is immediate to adapt the first part of the proof of Lemma 3.2 in order to see that $y \in C_{\psi_j,\kappa,d}$; in fact the whole statement of Lemma 3.2 holds because again the points $\zeta_{n/m}$ lie between two curves with an infinite order of tangency to the real axis.

We then follow the proof of Theorem 3.3 and obtain inequalities which are analogous to (3.3) but involve $\psi_j(m_\ell)$ instead of const $e^{-\alpha m_\ell}$. In order to conclude we only need to show that, for all j large enough and for all $\tau' > \tau$, there exist two positive constants c_0, c_1 such that

$$\forall N \geq 1, \quad \sum_{m=1}^{\infty} m^{\tau(N+2)} \psi_j(m) \leq c_0 c_1^N \Gamma(\tau'(N+2))$$

But this is an easy consequence of the fact that for all $\varepsilon > 0$ one has

$$\lim_{m\to+\infty} \exp(m^{1-\varepsilon})\psi_j(m) = 0$$

and, using the integral

$$\int_1^{+\infty} x^{\tau(N+2)} \exp(-x^{1-\varepsilon})dx \leq \frac{1}{1-\varepsilon}\Gamma\left(\frac{\tau(N+2)+\varepsilon}{1-\varepsilon}\right),$$

one can therefore bound the above series. □

5.3. An application to a nonlinear small divisor problem (semi-standard map)

In Section 4.2 we have studied the behaviour of the solution $f(q,z)$ of the linear equation

$$f(q,qz) - f(q,z) = g(z)$$

for q close to a resonance $\Lambda_0 = e^{2\pi i n_0/m_0}$. For q inside or outside the unit circle $\mathbb{S}^1$, the solution could be recovered from its asymptotic series via Borel-Laplace summation:

$$t f(\Lambda_0(1+t), z) = g \odot \mathcal{L}_{m_0} + \int_0^{\pm\infty} \hat{\Phi}(\xi, z)\, e^{-\xi/t}\, d\xi,$$

and the analytic continuation of the Borel transform $\hat{\Phi}$ w.r.t. ξ was carefully investigated.

We now indicate briefly that the same techniques can be adapted to a particular nonlinear equation. The reader is referred to a forthcoming paper for the proof of what follows. As for the motivation, the reader is referred to [BMS] where the connection between this nonlinear equation and the invariant circles of the Semi-Standard Map is explained.

We restrict ourselves to $\Lambda_0 = 1$ and inquire about the behaviour near that "resonance" of the solution $F(q,z)$ of the equation

$$F(q,qz) - 2F(q,z) + F(q,q^{-1}z) = -z\, e^{F(q,z)}. \tag{5.7}$$

There is an analytic solution F which, for each $q \in \mathbb{D} \cup \mathbb{E}$, is analytic for z close to the origin and which is characterized by $F(q,0) = 0$. It is shown in [BMS] that, as q tends non-tangentially to 1, $F(q,(q-1)^2 z)$ tends to $-2\log(1+z/2)$ (in that paper the non-tangential limit is computed for the other resonances as well). We now claim that this limit is nothing but the beginning of a Gevrey-1 asymptotic expansion and give some indications about the corresponding Borel transform.

We define the moving singular half-lines to be the half-lines $\pm\zeta_b(z)[1,+\infty[$ for $b \in \mathbb{Z}$, with

$$\zeta_b(z) = 2\pi(-i\log z + i\log 2 + \pi + 2\pi b).$$

THEOREM 5.4. *There is an analytic function $\hat{F}(\xi,z)$ which, for each $z \in \mathbb{D}_2$, is holomorphic for ξ in the complement of the half-lines $\pm\zeta_b(z)[1,+\infty[$ and has at most exponential growth on the lines passing through the origin and avoiding the points $\zeta_b(z)$, such that*

$$F(1+t, t^2 z) = -2\log(1+z/2) + \int_0^{\pm\infty} \hat{F}(\xi,z)\, e^{-\xi/t}\, d\xi.$$

In particular $F \in \mathcal{G}_1(1, zH^\infty(\mathbb{D}_r))$ for $0 < r < 2$.

The main difference with respect to the linear case is the necessity of rescaling the variable z when q approaches 1, instead of simply multiplying $f(q,z)$ by some regularizing factor like $t = q - 1$, and this is precisely due to the nonlinear character of (5.7). The analysis is of course more complicated, one needs to iterate a work which is analogous to that of Section 4.2, and this is why we restricted ourselves to the first resonance ($\Lambda_0 = 1$) and to the holomorphic star of $\hat{F}$ with respect to ξ. The case of the other resonances should be tractable. We suspect that $F(1+t, t^2 z)$ is resurgent with respect to t, i.e. that $\hat{F}(\xi, z)$ can be analytically continued with isolated singularities only, but this is probably much more difficult to prove.

APPENDIX

A.1. Hadamard's product

DEFINITION A1.1. The *Hadamard product* of two formal series

$$A(z) = \sum_{j \geq 0} a_j z^j , \quad B(z) = \sum_{j \geq 0} b_j z^j$$

is the formal series

$$(A \odot B)(z) = \sum_{j \geq 0} a_j b_j z^j .$$

If A and B are convergent power series with radii of convergence r_A and r_B then $A \odot B$ converges on the disk of radius $r_A r_B$.

We refer to [Be] for a detailed study of Hadamard algebras, i.e. algebras of formal power series in one variable with the product given by the Hadamard product.

The topological complex vector space $\mathbb{C}\{z\}$ with the product $\odot$ is a commutative complex algebra with unit $\delta(z) = \sum_{j=0}^{\infty} z^j$. The Hadamard product is a convolution: if $A, B \in \mathbb{C}\{z\}$ and γ is a simple continuous curve around the origin, contained in the convergence domain of A and B, one has

$$(A \odot B)(z) = \frac{1}{2\pi i} \int_\gamma A(w) B\left(\frac{z}{w}\right) \frac{dw}{w}$$

for $|z|$ small enough. The celebrated Hadamard Multiplication Theorem states that $A \odot B$ has in all sheets of its Riemann surface singularities at most at points lying over $\alpha \cdot \beta$, where α is a non-regular point of A and β is a non-regular point of B, and possibly at points lying over the origin [Sc]. A less general but more precise statement can be given as follows.

Let Ω be an open subset of $\mathbb{C}$ and let $\mathcal{O}(\Omega)$ denote the topological complex vector space of all functions which are holomorphic on Ω with the usual locally convex topology given by uniform convergence on compact subsets of Ω.

Let Ω_1, Ω_2 denote two open subset of $\mathbb{C}$ such that $0 \in \Omega_1 \cap \Omega_2$ and define

$$\Omega_1 \odot \Omega_2 = \mathbb{C} \setminus \{z \in \mathbb{C} \mid z = z_1 z_2,\ z_i \notin \Omega_i, i = 1, 2\}.$$

Then Hadamard's Theorem can be stated as follows ([Mü]):

THEOREM A1.1. *Let Ω_1, Ω_2 be as above, and let $L \in \mathcal{O}(\Omega_1)$. There exists a unique continuous linear mapping H_L from $\mathcal{O}(\Omega_1)$ into $\mathcal{O}(\Omega_1 \odot \Omega_2)$ such that, for all $\varphi \in \mathcal{O}(\Omega_2)$ and for all z with sufficiently small modulus, one has*

$$(H_L \varphi)(z) = (L \odot \varphi)(z).$$

In fact, we use mostly the case of functions analytic and bounded in disks, for which we have the following easy result (with the notation $B_r = zH^\infty(\mathbb{D}_r)$ for all $r > 0$):

LEMMA A1.1. *Let $0 < \rho' < \rho$ and $L \in B_{\rho'/\rho}$. The Hadamard product defines a bounded operator $\varphi \in B_\rho \mapsto L \odot \varphi \in B_{\rho'}$, whose operator norm is $\leq \|L\|_{B_{\rho'/\rho}}$.*

A.2. Some elementary properties of the fundamental solution

In this appendix we collect the statement and the proof of some elementary properties of the fundamental solution already used in the Introduction.

LEMMA A2.1. *Let $\delta = z(1-z)^{-1}$. If $q \in \mathbb{C}^* \setminus \mathcal{R}$, the series*

$$\sum_{\Lambda \in \mathcal{R}} \left(\frac{q}{\Lambda} - 1\right)^{-1} \mathcal{L}_{m(\Lambda)}$$

converges to $f_\delta(q, .)$ in $\mathbb{C}[[z]]$.

We recall that if J is a countable set and if $(f_j)_{j\in J}$ is a family of formal series, this family is summable if for all integer m, the set $\{j \in J \mid f_j \notin O(z^m)\}$ is finite. In this case the series $\sum_{j\in J} f_j$ is called convergent in $\mathbb{C}[[z]]$ and its sum is a formal series independent on the choice of an ordering on J. (This is the well-known notion of convergence associated to the z-adic valuation).

PROOF. The valuation of $\mathcal{L}_{m(\Lambda)}$ is $m(\Lambda)$ and for each m the set $\mathcal{R}_m^*$ is finite. The series f mentioned in the above lemma converges thus formally, and it can be rewritten as

$$f = \sum_{m\geq 1} \sum_{\Lambda \in \mathcal{R}_m^*} \sum_{j \geq 1} \frac{z^{jm}}{jm} \left(\frac{q}{\Lambda} - 1\right)^{-1} = \sum_{(m,j)\in \mathbb{N}^* \times \mathbb{N}^*} \sum_{\Lambda \in \mathcal{R}_m^*} \frac{z^{jm}}{jm} \left(\frac{q}{\Lambda} - 1\right)^{-1}.$$

By reordering the terms of the summable family indexed by $\mathbb{N}^* \times \mathbb{N}^*$, one finds

$$f = \sum_{\ell \geq 1} \sum_{m|\ell} \sum_{\Lambda \in \mathcal{R}_m^*} \frac{z^\ell}{\ell} \left(\frac{q}{\Lambda} - 1\right)^{-1} = \sum_{\ell\geq 1} \frac{z^\ell}{\ell} \sum_{\Lambda\in\mathcal{R}_\ell} \left(\frac{q}{\Lambda} - 1\right)^{-1}.$$

In the coefficient of z^ℓ one recognizes the decomposition into simple elements of the corresponding coefficient in

$$f_\delta(q, z) = \sum_{\ell \geq 1} \frac{z^\ell}{q^\ell - 1}.$$

□

By means of the Hadamard product, the "decomposition into simple elements" just proved for the fundamental solution can be extended to the general solution f_g of (1.1):

PROPOSITION A2.1. *Let $g \in z\mathbb{C}[[z]]$ and $q \in \mathbb{C}^* \setminus \mathcal{R}$. The solution $f_g(q, .)$ can be written as the sum in $\mathbb{C}[[z]]$ of the series*

$$\sum_{\Lambda \in \mathcal{R}} \left(\frac{q}{\Lambda} - 1\right)^{-1} g \odot \mathcal{L}_{m(\Lambda)}.$$

PROOF. The identities

$$g = g \odot \delta , \quad f_g = g \odot f_\delta$$

are evident. On the other hand, for any summable family $(f_j)_{j\in J}$ de $\mathbb{C}[[z]]$, the family $(g \odot f_j)_{j\in J}$ is summable (because the Hadamard product with a formal series g does not decrease the valuation), and

$$g \odot \sum_{j\in J} f_j = \sum_{j\in J} g \odot f_j .$$

The result follows then from Lemma A2.1. □

LEMMA A2.2. *Let*

$$S = \{ q = e^{2\pi i x} \mid x \in \mathbb{R} \setminus \mathbb{Q}, \ \limsup_{k\to\infty} \frac{\log m_{k+1}}{m_k} = +\infty \}.$$

where $(n_k/m_k)_{k\geq 1}$ is the sequence of the convergents to x (see Appendix A.3 for its definition and properties). For each $q \in S$ the fundamental solution $f_\delta(q,z) = \sum_{n\geq 1} \frac{z^n}{q^n-1}$ diverges. The set S is a G_δ-dense subset of $\mathbb{S}^1$ of measure zero. On the contrary, if $q = e^{2\pi i x}$ and $\limsup_{k\to\infty} \frac{\log m_{k+1}}{m_k} \leq M$, then $f_\delta(q,z)$ converges in the disk $|z| < e^{-M}$.

PROOF. The divergence of f_δ when $q \in S$ is well-known ([HL], [Sim]), together with the convergence statement. S is a G_δ-dense in $\mathbb{S}^1$, since it is immediate to check that

$$S = \bigcap_{j\geq 0} \bigcup_{n/m} \{ q = e^{2\pi i x} \mid |x - n/m| < \frac{e^{-jm}}{m} \}.$$

This also shows that S has measure zero. □

A.3. Some arithmetical results. Continued fractions

We gather here a few facts on continued fractions that we use in Chapters 2, 3 and Appendices A.2, A.4. We refer the reader to [HW], [Khi] and [MMY] for more details.

A.3.1. Let $[x]$ denote the usual integer part of a real number x, $\{x\}$ its fractional part: $\{x\} = x - [x]$.

To each $x \in \mathbb{R} \setminus \mathbb{Q}$ we associate its continued fraction expansion as follows. Let

$$x_0 = x - [x], \quad a_0 = [x],$$

then one obviously has $x = a_0 + x_0$, $a_0 \in \mathbb{Z}$, $x_0 \in]0,1[$. We consider the iteration of the Gauss map $A :]0,1[\to [0,1[$, $A(x) = \{x^{-1}\}$, and we define inductively

$$x_{k+1} = \left\{ \frac{1}{x_k} \right\}, \quad a_{k+1} = \left[\frac{1}{x_k} \right].$$

This can be done for all $k \geq 0$ since x is irrational, thus

$$x_k^{-1} = a_{k+1} + x_{k+1}, \qquad x_{k+1} \in]0,1[, \qquad a_{k+1} \in \mathbb{N}^*,$$

and we have

$$x = a_0 + x_0 = a_0 + \cfrac{1}{a_1 + x_1} = \ldots = a_0 + \cfrac{1}{a_1 + \cfrac{1}{a_2 + \ddots + \cfrac{1}{a_k + x_k}}}.$$

We will write

$$x = [a_0, a_1, \ldots, a_k, \ldots].$$

The integers $a_0, a_1, \ldots, a_k, \ldots$ are called the *partial quotients* of x. The kth *convergent* is defined by

$$\frac{n_k}{m_k} = [a_0, a_1, \ldots, a_k] = a_0 + \cfrac{1}{a_1 + \cfrac{1}{a_2 + \ddots + \cfrac{1}{a_k}}}$$

and $\frac{n_k}{m_k} \to x$ as $k \to \infty$. It is immediate to check that the numerators n_k and denominators m_k are recursively determined by

$$\text{(A3.0)} \qquad \begin{array}{lll} n_{-2} = 0, & n_{-1} = 1, & n_k = a_k n_{k-1} + n_{k-2}, \\ m_{-2} = 1, & m_{-1} = 0, & m_k = a_k m_{k-1} + m_{k-2}, \end{array}$$

for $k \geq 0$. Observe that $x_k = [0, a_{k+1}, a_{k+2}, \ldots]$. Moreover

$$\text{(A3.1)} \qquad x = \frac{n_k + n_{k-1} x_k}{m_k + m_{k-1} x_k}$$

$$\text{(A3.2)} \qquad x_k = -\frac{m_k x - n_k}{m_{k-1} x - n_{k-1}}$$

$$\text{(A3.3)} \qquad m_k n_{k-1} - n_k m_{k-1} = (-1)^k.$$

Let

$$\beta_k = \Pi_{i=0}^k x_i = (-1)^k (m_k x - n_k) \quad \text{for } k \geq 0, \quad \text{and } \beta_{-1} = 1.$$

Then

$$x_k = \frac{\beta_k}{\beta_{k-1}}$$

$$\beta_{k-2} = a_k \beta_{k-1} + \beta_k.$$

A.3.2. From the definitions given above, one easily proves by induction the following proposition (we refer, for example, to [MMY] for its proof):

PROPOSITION A3.1. *For all $x \in \mathbb{R} \setminus \mathbb{Q}$ and for all $k \geq 0$ one has*

(i) $m_{k+2} > m_{k+1} > 0$;

(ii) $n_k > 0$ *when* $x > 0$ *and* $n_k < 0$ *when* $x < 0$;

(iii) $|m_k x - n_k| = \dfrac{1}{m_{k+1} + m_k x_{k+1}}$, *so that* $\frac{1}{2} < \beta_k m_{k+1} < 1$;

(iv) $\beta_k \leq g^k$, *where* $g = \frac{\sqrt{5}-1}{2}$.

REMARK A3.1. Note that from *(iii)* and *(iv)* one gets $m_k \geq \frac{1}{2} G^{k-1}$ with $G = g^{-1} = \frac{\sqrt{5}+1}{2}$.

REMARK A3.2. From *(iii)* one gets

$$\text{(A3.4)} \qquad \frac{1}{2m_k m_{k+1}} < \frac{1}{m_k(m_k + m_{k+1})} < \left| x - \frac{n_k}{m_k} \right| < \frac{1}{m_k m_{k+1}}$$

Note also that *(iv)* remains valid for $x \in \mathbb{Q}$: in this case there exists $j \geq 0$ such that $x_j = 0$ and the x_k with $k \geq j$ are undefined; we set $\beta_k = 0$ for all $k \geq j$.

A partial converse of *(iii)*, Proposition A3.1, is provided by the following very useful Proposition (see [HW], Theorem 184, p. 153):

PROPOSITION A3.2. *Let* $x \in \mathbb{R} \setminus \mathbb{Q}$. *If* $\left|\frac{n}{m} - x\right| < \frac{1}{2m^2}$ *then* $\frac{n}{m}$ *is a convergent of* x.

The bound *(iii)*, Proposition A3.1, on the approximation provided by the convergents implies that $m_k \left|m_k x - n_k\right| < a_{k+1}^{-1}$. One can also prove the following ([HW], Theorem 193, p. 164)

PROPOSITION A3.3. *For each* $x \in \mathbb{R} \setminus \mathbb{Q}$, *there exist infinitely many rational numbers* $\frac{n}{m}$ *such that* $\left|\frac{n}{m} - x\right| < \frac{1}{\sqrt{5}\, m^2}$.

The following property will be used in Appendix A.4.1:

LEMMA A3.1. *For all* $k \geq 1$,

$$m_k^2 \left| x - \frac{n_k}{m_k} \right| = \frac{1}{x_k' + y_k}$$

with $x_k' = [a_{k+1}, a_{k+2}, \ldots]$ *and* $y_k = \frac{m_{k-1}}{m_k}$.

PROOF. Rewrite the identity in *(iii)*, Proposition A3.1, using $x_k' = a_{k+1} + x_{k+1}$ and $m_{k+1} - a_{k+1} m_k = m_{k-1}$. □

Among all rational approximations the convergents are the most accurate in a very precise sense:

PROPOSITION A3.4 (The law of best approximation). *If* $k \geq 1$ *and* n, m *are integers such that* $1 \leq m \leq m_k$ *but* $(n, m) \neq (n_k, m_k)$, *then* $|mx - n| > |m_k x - n_k|$. *Moreover, if* $(n, m) \neq (n_{k-1}, m_{k-1})$ *and* $k > 1$, *then* $|mx - n| > |m_{k-1} x - n_{k-1}|$.

For a proof see [HW], Theorem 182, p. 151–52.

A.3.3. According to Definition 3.5, x is Diophantine with exponent 2, i.e. $x \in \mathrm{DC}_2$, iff the quantities $m^2 \left|\frac{n}{m} - x\right|$ are bounded from below. In view of Proposition A3.2, it is sufficient to require this property for the convergents (since $m^2 \left|\frac{n}{m} - x\right| \geq \frac{1}{2}$ when $\frac{n}{m}$ is not a convergent). From the inequalities (A3.4), it follows easily that

$$x \in \mathrm{DC}_2 \;\Leftrightarrow\; \frac{m_{k+1}}{m_k} \text{ bounded.}$$

Constant-type numbers are usually defined as the numbers whose partial quotients are bounded. If $M = \sup\{a_k\} < \infty$,

$$\text{(A3.5)} \qquad \forall k \geq 0, \quad \frac{m_{k+1}}{m_k} = \frac{a_k m_k + m_{k-1}}{m_k} < a_k + 1 \leq M + 1.$$

Conversely $a_k < \frac{m_{k+1}}{m_k}$, hence

$$\frac{m_{k+1}}{m_k} \text{ bounded} \;\Leftrightarrow\; x \text{ has constant type.}$$

Notice that, in that situation, $m_{k-2} \geq \frac{1}{M+1} m_{k-1}$ (with $k \geq 2$) implies that $m_k = a_k m_{k-1} + m_{k-2} \geq m_{k-1} + \frac{1}{M+1} m_{k-1}$, that is

$$\forall k \geq 2, \quad \frac{m_k}{m_{k-1}} \geq 1 + \frac{1}{M+1}. \tag{A3.6}$$

A particular case of constant-type number is obtained when the continued fraction expansion is eventually periodic, i.e. when there exist $L \geq 0$ and $K \geq 1$ such that $a_{k+K} = a_k$ for all $k \geq L$. Such numbers are characterized by the celebrated Lagrange's theorem ([HW], Theorems 176 and 177):

THEOREM A3.1. *The number x is algebraic of degree 2 iff its continued fraction expansion is eventually periodic.*

A.4. Proof of Lemma 3.3

A.4.1. Let $\alpha \in]0,1[$ be a quadratic irrational number. Recall that $\mathbb{N}^* \times \mathbb{Z}$ has been partitioned into

$$\mathcal{E}^- = \{ (D,N) \in \mathbb{N}^* \times \mathbb{Z} \mid N/D < \alpha \} \quad \text{and} \quad \mathcal{E}^+ = \{ (D,N) \in \mathbb{N}^* \times \mathbb{Z} \mid N/D > \alpha \}.$$

Our aim is to find numbers $\nu_+, \nu_-, \nu'_+, \nu'_-$ with $\nu'_\pm > \nu_\pm$, and decompositions

$$\mathcal{E}^+ = \mathcal{F}^+ \cup \mathcal{E}^+_* \cup \mathcal{A}^+, \quad \mathcal{E}^- = \mathcal{F}^- \cup \mathcal{E}^-_* \cup \mathcal{A}^-,$$

with $\mathcal{F}^\pm$ finite, satisfying specific properties about the quantities $D^2 \left|\frac{N}{D} - \alpha\right|$ corresponding to the members of $\mathcal{E}^\pm_*$ and $\mathcal{A}^\pm$: it is required that $D^2 \left|\frac{N}{D} - \alpha\right| \geq \nu'_\pm$ for all $(N,D) \in \mathcal{E}^\pm_*$, whereas $\mathcal{A}^\pm$ should consist of an infinite sequence $\{(D^\pm_p, N^\pm_p)\}$ for which $D^2 \left|\frac{N}{D} - \alpha\right| = \nu_\pm + O(D^{-2})$. Lastly, $\{D^\pm_p\}$ is required to be increasing, with $D^\pm_{p+1}/D^\pm_p$ bounded and $\sum (D^\pm_p)^{-1/2}$ convergent.

Notice that the above properties will imply

$$\nu_\pm = \liminf_{(D,N)\in\mathcal{E}^\pm} D^2 \left| \frac{N}{D} - \alpha \right|,$$

i.e. $\nu_\pm = \nu_\pm(e^{2\pi i \alpha})$ with the notation of Definition 3.6. Hence all the conclusions of Lemma 3.3 will hold if we choose $\kappa'_\pm = (\nu'_\pm)^{1/2}$.

We will construct $\mathcal{A}^\pm$ by extracting appropriate subsequences from the sequence $\{\frac{n_k}{m_k}\}$ of the convergents of α. Convergents are good candidates for the elements of $\mathcal{A}^\pm$ because

$$\forall k \geq 0, \quad m_k^2 \left| \frac{n_k}{m_k} - \alpha \right| < 1 \tag{A4.1}$$

by virtue of (A3.4), and they fall alternately in $\mathcal{E}^-$ and $\mathcal{E}^+$ according to the parity of k because of (A3.2). Then we will have to define $\nu'_\pm$ and to distribute the rational numbers other than the convergents between $\mathcal{F}^\pm$ and $\mathcal{E}^\pm_*$.

Let $P(X)$ be the polynomial of definition of α:

$$P(X) = aX^2 + bX + c = a(X-\alpha)(X-\overline{\alpha}), \qquad a,b,c \in \mathbb{Z}, \qquad a \geq 1,$$

$$\alpha = \frac{-b + \varepsilon\sqrt{\Delta}}{2a}, \quad \overline{\alpha} = \frac{-b - \varepsilon\sqrt{\Delta}}{2a}, \quad \varepsilon \in \{-1,+1\}, \quad \Delta = b^2 - 4ac \geq 2.$$

For all $(D,N) \in \mathbb{N}^* \times \mathbb{Z}$ the expression

$$F(N,D) = aN^2 + bND + cD^2 = a\left(\frac{N}{D} - \overline{\alpha}\right)\left(\frac{N}{D} - \alpha\right) D^2$$

can assume only nonzero integral values and will allow us to control the quantity $\left|\frac{N}{D} - \alpha\right| D^2$ when it is small.

A.4.2. By Lagrange's theorem, the continued fraction expansion of α is eventually periodic; we denote it by

$$\alpha = [a_0, a_1, \dots, a_{L-1}, \overline{a_L, \dots, a_{L+K-1}}],$$

where $K \geq 1$ is the period, $L \in \mathbb{N}$, and the line above a finite sequence of coefficients indicates its repetition in the continued fraction. The periodicity of the continued fraction expansion of α will reflect somehow on the polynomial $F(X,Y) = Y^2 P(\frac{X}{Y})$ when evaluated on the convergents:

LEMMA A4.1. *For all $k \geq L$,*

$$F(n_k, m_k) = (-1)^K F(n_{K+k}, m_{K+k}).$$

PROOF. Let us first treat the case where $L = 0$. We recall that

$$(n_{-2}, m_{-2}) = (0,1), \quad (n_{-1}, m_{-1}) = (1,0),$$
$$\forall k \geq 0, \quad (n_k, m_k) = a_k(n_{k-1}, m_{k-1}) + (n_{k-2}, m_{k-2}).$$

On the one hand the periodicity property $a_{k+K} = a_k$ allows one to check easily (by induction on k) the following relationship between (n_{k+K}, m_{k+K}) and (n_k, m_k):

$$\text{(A4.2)} \qquad \forall k \geq -2, \quad (n_{K+k}, m_{K+k}) = n_k(n_{K-1}, m_{K-1}) + m_k(n_{K-2}, m_{K-2}).$$

On the other hand, still by virtue of the periodicity of the continued fraction,

$$\alpha = [a_0, a_1, \dots, a_{K-1}, \alpha] = \frac{\alpha n_{K-1} + n_{K-2}}{\alpha m_{K-1} + m_{K-2}}$$

(the last equality follows from (A3.1) with $x_{K-1} = \frac{1}{\alpha}$). Hence the polynomial

$$P_1(X) = m_{K-1}X^2 + (m_{K-2} - n_{K-1})X - n_{K-2}$$

vanishes at $X = \alpha$, i.e. belongs to the ideal of $\mathbb{Q}[X]$ generated by $P(X)$:

$$P_1(X) = \frac{m_{K-1}}{a} P(X).$$

We can thus content ourselves with checking that

$$\forall k \geq 0, \quad F_1(n_{K+k}, m_{K+k}) = (-1)^K F_1(n_k, m_k),$$

where $F_1(X,Y) = Y^2 P_1(\frac{X}{Y}) = m_{K-1}X^2 + (m_{K-2} - n_{K-1})XY - n_{K-2}Y^2$.

This is a simple computation: for $k \geq -2$, using (A4.2),

$$\begin{aligned} F_1(n_{K+k}, m_{K+k}) &= m_{K-1}(n_k n_{K-1} + m_k n_{K-2})^2 \\ &\quad + (m_{K-2} - n_{K-1})(n_k n_{K-1} + m_k n_{K-2})(n_k m_{K-1} + m_k m_{K-2}) \\ &\quad - n_{K-2}(n_k m_{K-1} + m_k m_{K-2})^2 \\ &= A n_k^2 + B n_k m_k + C m_k^2, \end{aligned}$$

with $A = m_{K-1}(n_{K-1} m_{K-2} - m_{K-1} n_{K-2}) = (-1)^K m_{K-1}$,

$B = (m_{K-2} - n_{K-1})(n_{K-1} m_{K-2} - m_{K-1} n_{K-2}) = (-1)^K (m_{K-2} - n_{K-1})$,

$C = n_{K-2}(m_{K-1} n_{K-2} - n_{K-1} m_{K-2}) = (-1)^{K-1} n_{K-2}$

(thanks to (A3.3)). This ends the proof of Lemma A4.4 in the case $L = 0$.

We now proceed by induction on L. We suppose that

$$\alpha = [a_0, a_1, \dots, a_{L-1}, \overline{a_L, \dots, a_{L+K-1}}]$$

with $L \geq 1$, and that the convergents $\{n'_k / m'_k\}$ of

$$\alpha' = [a_1, a_2, \dots, a_{L-1}, \overline{a_L, \dots, a_{L+K-1}}]$$

satisfy

$$\forall k \geq L-1, \quad G(n'_k, m'_k) = (-1)^K G(n'_{K+k}, m'_{K+k}),$$

where $G(X,Y) = Y^2 Q(\frac{X}{Y})$ and $Q(X)$ is the polynomial of definition of α'.

The identity

$$\alpha = [a_0, \alpha'] = a_0 + \frac{1}{\alpha'}$$

shows that the polynomial

$$P_1(X) = (X - a_0)^2 Q\Big(\frac{1}{X - a_0}\Big) \in \mathbb{Z}[X]$$

vanishes at $X = \alpha$, thus $P_1(X)$ is a rational multiple of the polynomial of definition of α and we can content ourselves with checking that

$$\forall k \geq L, \quad F_1(n_{K+k}, m_{K+k}) = (-1)^K F_1(n_k, m_k),$$

where $F_1(X,Y) = Y^2 P_1(\frac{X}{Y}) = (X - a_0 Y)^2 Q\big(\frac{Y}{X - a_0 Y}\big) = G(Y, X - a_0 Y)$.

Let us express the convergents of α in terms of those of α': if $k \geq 1$,

$$\frac{n_k}{m_k} = [a_0, a_1, \dots, a_k] = a_0 + \frac{1}{[a_1, a_2, \dots, a_k]} = a_0 + \frac{m'_{k-1}}{n'_{k-1}},$$

thus $n_k = a_0 n'_{k-1} + m'_{k-1}$, $m_k = n'_{k-1}$ and $n_k - a_0 m_k = m'_{k-1}$. Hence

$$\forall k \geq 1, \quad F_1(n_k, m_k) = G(n'_{k-1}, m'_{k-1}),$$

and by the inductive hypothesis

$$\forall k \geq L, \quad F_1(n_{K+k}, m_{K+k}) = (-1)^K F_1(n_k, m_k).$$

□

A.4.3. We recall that $(-1)^k(x - \frac{n_k}{m_k}) > 0$, thus

$$(m_k, n_k) \in \mathcal{E}^+ \text{ if } k \text{ is odd}, \quad (m_k, n_k) \in \mathcal{E}^- \text{ if } k \text{ is even}.$$

As a consequence of the previous lemma, the sets $\mathcal{R}^+ = \{\, |F(n_k, m_k)|,\ k \text{ odd} \geq L \}$ and $\mathcal{R}^- = \{\, |F(n_k, m_k)|,\ k \text{ even} \geq L \}$ are finite. We define

$$r^+ = \min \mathcal{R}^+, \quad r^- = \min \mathcal{R}^-.$$

If moreover $\mathcal{R}^\pm$ is not reduced to a single element, we define $r_*^\pm = \min \mathcal{R}^\pm \setminus \{r^\pm\}$. Since

$$|F(n_k, m_k)| = a \left| \frac{n_k}{m_k} - \overline{\alpha} \right| \left| \frac{n_k}{m_k} - \alpha \right| m_k^2 \tag{A4.3}$$

and $a \left| \frac{n_k}{m_k} - \overline{\alpha} \right|$ tends to the limit $a\,|\alpha - \overline{\alpha}| = \sqrt{\Delta}$ as $k \to \infty$, the inequality (A4.1) implies that $\mathcal{R}^+$ and $\mathcal{R}^-$ are bounded by that limit. We define

$$\nu_\pm = \frac{r^\pm}{\sqrt{\Delta}} \leq 1.$$

Let $k_0 > L$ be large enough so that

$$\forall k \geq k_0, \quad \left| \frac{n_k}{m_k} - \alpha \right| \leq C, \tag{A4.4}$$

where the positive number C will be specified later, in Formula (A4.7). We define

$$\mathcal{A}^+ = \{\, (m_k, n_k),\ k > k_0 \text{ odd such that } |F(n_k, m_k)| = r^+ \},$$
$$\mathcal{A}^- = \{\, (m_k, n_k),\ k > k_0 \text{ even such that } |F(n_k, m_k)| = r^- \}.$$

The members of $\mathcal{A}^\pm$ can be enumerated as

$$\mathcal{A}^\pm = \{\, (D_p^\pm, N_p^\pm),\ p \geq 0 \}, \quad (D_p^\pm, N_p^\pm) = (m_{k^\pm(p)}, n_{k^\pm(p)}),$$

with an increasing sequence $k^\pm(p)$ of odd/even integers larger than k_0. Lemma A4.1 then ensures that, whenever $(m_k, n_k) \in \mathcal{A}^\pm$, we have also $(m_{k+2K}, n_{k+2K}) \in \mathcal{A}^\pm$, hence

$$k^\pm(p+1) \leq k^\pm(p) + 2K$$

(in the previous sentence, $2K$ can be replaced by K if K is even; we only need to avoid a jump from one of the sets $\mathcal{E}^+, \mathcal{E}^-$ to the other).

A.4.4. Let $(D, N) \in \mathcal{A}^\pm$. Since C will not exceed $\frac{|\alpha - \overline{\alpha}|}{9}$ (see Formula (A4.7) below), the reals $\frac{N}{D} - \overline{\alpha}$ and $\alpha - \overline{\alpha}$ have same sign and, using (A4.3), we can easily compute the discrepancy between $D^2 \left| \frac{N}{D} - \alpha \right| = \dfrac{r^\pm}{a \left| \frac{N}{D} - \overline{\alpha} \right|}$ and $\nu^\pm = \frac{r^\pm}{a|\alpha - \overline{\alpha}|}$:

$$D^2 \left| \frac{N}{D} - \alpha \right| - \nu_\pm = \frac{r^\pm}{a|\alpha - \overline{\alpha}|} \cdot \frac{\left| \alpha - \frac{N}{D} \right|}{\left| \frac{N}{D} - \overline{\alpha} \right|} = \frac{1}{|\alpha - \overline{\alpha}|} \left(\frac{r^\pm}{a \left| \frac{N}{D} - \overline{\alpha} \right|} \right)^2 \frac{1}{D^2} = O(D^{-2}).$$

The convergence of the series $\sum (D_p^\pm)^{-1/2}$ is guaranteed by Remark A3.1: its terms are part of the series $\sum m_k^{-1/2}$ which is convergent.

Let us check that the sequences D^+_{p+1}/D^+_p and D^-_{p+1}/D^-_p are bounded. We introduce

$$M = \max\{a_k\} \geq 1 \tag{A4.5}$$

(recall that there are only a finite number of quotients a_k, by Lagrange's theorem) and we recall that the numbers $\frac{m_{k+1}}{m_k}$ are bounded by $M+1$ according to (A3.5). The conclusion follows from the remark on the sequences $k^+(p)$ and $k^-(p)$ at the end of Paragraph A.4.3:

$$\frac{D^{\pm}_{p+1}}{D^{\pm}_p} = \frac{m_{k^{\pm}(p+1)}}{m_{k^{\pm}(p)}} \leq \frac{m_{k^{\pm}(p)+2K}}{m_{k^{\pm}(p)}} < (M+1)^{2K}.$$

A.4.5. There remains only to find $\nu'_{\pm}$, $\mathcal{F}^{\pm}$, $\mathcal{E}^{\pm}_*$ with the required properties. Let $C_0 = \frac{1}{M+1}$; we define ν'_+ and ν'_- by

$$\nu'_{\pm} = \min\left\{\frac{r^{\pm}+r^{\pm}_*}{2\sqrt{\Delta}}, 1+C_0^2, \frac{1+2C_0}{1+C_0}, \left(1+\frac{C_0}{4}\right)\nu_{\pm}\right\}. \tag{A4.6}$$

Of course, it is understood that if $\mathcal{R}^+$ or $\mathcal{R}^-$ is reduced to a single element, in which case r^+_* or r^-_* is not defined, the above minimum is taken only over the last three numbers of the right-hand side. Since $0 < C_0 \leq \frac{1}{2}$, $\nu_{\pm} = \frac{r^{\pm}}{\sqrt{\Delta}} \leq 1$ and $r^{\pm}_* > r^{\pm}$ when it is defined, we always have $\nu'_{\pm} > \nu_{\pm}$.

We also define now the number C upon which k_0 depends through (A4.4):

$$C = |\alpha - \overline{\alpha}| \cdot \min\left\{\frac{1}{1+\frac{4}{C_0}}, \frac{r^+_* - r^+}{r^+_* + r^+}, \frac{r^-_* - r^-}{r^-_* + r^-}\right\} \tag{A4.7}$$

(with the same convention as previously if r^+_* or r^-_* is undefined), and the finite sets $\mathcal{F}^+$ and $\mathcal{F}^-$:

$$\mathcal{F}^{\pm} = \left\{(D,N) \in \mathcal{E}^{\pm} \mid D \leq m_{k_0} \text{ and } \alpha D - \frac{\nu'_-}{D} < N < \alpha D + \frac{\nu'_+}{D}\right\}$$

(where $[\,.\,]$ denotes the integer part of a real number).

We observe that each member (D,N) of the complement set $\mathcal{E}^{\pm}_* = \mathcal{E}^{\pm} \setminus (\mathcal{A}^{\pm} \cup \mathcal{F}^{\pm})$ necessarily falls in one of three distinct categories:

(i) either $D \leq m_{k_0}$, and $N \geq \alpha D + \frac{\nu'_+}{D}$ ('+' case) or $N \leq \alpha D - \frac{\nu'_-}{D}$ ('−' case);
(ii) or $(D,N) = (m_k, n_k)$ with $k > k_0$ and $|F(n_k, m_k)| \geq r^{\pm}_*$;
(iii) or $D > m_{k_0}$ and N/D is not one of the convergents of α.

The inequality

$$D^2\left|\frac{N}{D} - \alpha\right| \geq \nu'_{\pm} \tag{A4.8}$$

is obvious in the first case. In the second case it follows from (A4.3), which implies

$$m_k^2\left|\frac{n_k}{m_k} - \alpha\right| \geq \frac{r^{\pm}_*}{\sqrt{\Delta}} \cdot \frac{|\alpha - \overline{\alpha}|}{\left|\frac{n_k}{m_k} - \overline{\alpha}\right|},$$

and from the inequality

$$r^{\pm}_*|\alpha - \overline{\alpha}| \geq \frac{r^{\pm}_* + r^{\pm}}{2}\left|\frac{n_k}{m_k} - \overline{\alpha}\right|$$

which is a consequence of (A4.4) and of the condition $C \le \frac{r_*^\pm - r^\pm}{r_*^\pm + r^\pm}|\alpha - \overline{\alpha}|$ imposed by (A4.7) in that case. The next two paragraphs are devoted to the proof of (A4.8) in the third case.

A.4.6. We are left with the case of an element (D,N) of $\mathcal{E}^\pm$ such that N/D is not a convergent of α and $D > m_{k_0}$. Let $k > k_0$ be the unique integer such that $m_{k-1} \le D < m_k$. We will suppose k odd for the sake of clarity, just to have $\frac{n_{k-1}}{m_{k-1}}, \frac{n_{k+1}}{m_{k+1}}, \alpha, \frac{n_k}{m_k}$ in that order on the real line (we leave it to the reader to adapt the following arguments to the case of even k).

Let I_k denote the closed interval $\left[\frac{n_{k-1}}{m_{k-1}}, \frac{n_k}{m_k}\right]$, whose length is exactly $\frac{1}{m_{k-1}m_k}$ (because of (A3.3)). Since $m_{k-1}N - n_{k-1}D$ is a nonzero integer,

$$\left|\frac{N}{D} - \frac{n_{k-1}}{m_{k-1}}\right| \ge \frac{1}{m_{k-1}D} > \operatorname{length}(I_k),$$

thus $\frac{N}{D}$ is either to the left or to the right of I_k.

In the first case, necessarily $(D,N) \in \mathcal{E}^-$ and

$$\begin{aligned}\left|\frac{N}{D} - \alpha\right| &> \left|\frac{N}{D} - \frac{n_{k-1}}{m_{k-1}}\right| + \left|\frac{n_{k-1}}{m_{k-1}} - \frac{n_{k+1}}{m_{k+1}}\right| \\ &\ge \frac{1}{m_{k-1}D} + \frac{1}{m_{k-1}m_{k+1}} \ge \frac{1+C_0^2}{D^2},\end{aligned}$$

using $m_{k-1} \le D$, $m_{k+1} \le (M+1)^2 D$ (which follows from (A3.5)) and $C_0 = \frac{1}{M+1}$. The obtained inequality yields (A4.8) since $1 + C_0^2 \ge \nu'_-$.

Let us assume we are in the second case: $\frac{N}{D} > \frac{n_k}{m_k}$. In particular $(D,N) \in \mathcal{E}^+$. We will distinguish two subcases, according to the value of the positive integer $m_{k-1}N - n_{k-1}D$. If this number is not 1,

$$\begin{aligned}\frac{N}{D} - \alpha > \frac{N}{D} - \frac{n_k}{m_k} &= \left(\frac{N}{D} - \frac{n_{k-1}}{m_{k-1}}\right) - \left(\frac{n_k}{m_k} - \frac{n_{k-1}}{m_{k-1}}\right) \\ &\ge \frac{2}{m_{k-1}D} - \frac{1}{m_{k-1}m_k} = \frac{\Phi}{D^2}.\end{aligned}$$

We can write Φ as a polynomial function of degree 2 of a real variable $\lambda = \frac{D}{m_{k-1}}$ ranging between 1 and $\Lambda = \frac{m_k}{m_{k-1}}$:

$$\Phi = \Phi(\lambda) = \lambda\left(2 - \frac{\lambda}{\Lambda}\right).$$

Since $\frac{d\Phi}{d\lambda}$ is decreasing, we have $\Phi(\lambda) \ge \min\{\Phi(1), \Phi(\Lambda)\}$. But $\Phi(1) = \frac{1+2C_0}{1+C_0} \ge \nu'_+$, and $\Phi(\Lambda) = \Lambda \ge 1 + C_0$ by (A3.6), which is also $\ge \nu'_+$. Hence (A4.8) is true in the first subcase.

A.4.7. There remains only to prove (A4.8) in the last subcase: $\frac{N}{D}$ lies to the right of α and is not a convergent, $m_{k-1} \le D < m_k$ and $m_{k-1}N - n_{k-1}D = 1$, where k is odd and greater than k_0.

LEMMA A4.2. *With these hypotheses,*

$$D^2\left(\frac{N}{D} - \alpha\right) \ge \left(1 + \frac{C_0}{2}\right) \cdot m_k^2 \left(\frac{n_k}{m_k} - \alpha\right). \tag{A4.9}$$

PROOF. Thanks to (A3.3), we can write

$$D = m_k - r m_{k-1}, \quad N = n_k - r n_{k-1},$$

for some integer $r \geq 1$. Let

$$y_k = \frac{m_{k-1}}{m_k}.$$

Since $\frac{m_{k-1}}{m_k} \leq \frac{D}{m_k}$, we get $y_k \leq 1 - r y_k$, thus

$$1 \leq r \leq \frac{1}{y_k} - 1. \tag{A4.10}$$

Notice also that $y_k \leq \frac{1}{1+r} \leq \frac{1}{2}$.

Let us compute the left-hand side of (A4.9) in terms of m_k, n_k, y_k and r: using (A3.3) we have $-n_{k-1} = \frac{1}{m_k} - y_k n_k$, thus

$$N = n_k(1 - r y_k) + \frac{r}{m_k},$$

and $D = m_k(1 - r y_k)$, therefore

$$D(N - D\alpha) = m_k(n_k - m_k\alpha)(1 - r y_k)^2 + r(1 - r y_k).$$

Using the notations of Lemma A3.1, we have $m_k(n_k - m_k\alpha) = \frac{1}{x'_k + y_k}$ with

$$x'_k = a_{k+1} + \cfrac{1}{a_{k+2} + \cfrac{1}{a_{k+3} + \ddots}} \geq 1 + C_0$$

(the inequality follows from $a_{k+1}, a_{k+3} \geq 1$ and $a_{k+2} \leq M$). Hence

$$D(N - D\alpha) = \frac{(1 - r y_k)(1 + r x'_k)}{x'_k + y_k} = m_k(n_k - m_k\alpha)\Psi, \qquad \Psi = (1 - r y_k)(1 + r x'_k).$$

To conclude, we need only to check that $\Psi \geq 1 + \frac{C_0}{2}$.

But Ψ may be viewed as a polynomial function of degree 2 of a real variable r, whose minimum over the range (A4.10) must be attained at one extremity of this range (because the derivative is decreasing). When $r = 1$, $\Psi = (1 - y_k)(1 + x'_k) \geq (1+x'_k)/2 \geq 1 + \frac{C_0}{2}$. And when $r = \frac{1}{y_k} - 1$, $\Psi = y_k(1 + \frac{x'_k}{y_k} - x'_k) = 1 + (x'_k - 1)(1 - y_k) \geq 1 + \frac{C_0}{2}$. Thus Ψ is always bounded from below as required in that range. □

Inequality (A4.8) (with a '+' sign) follows from (A4.9). We have indeed, by (A4.3),

$$m_k^2\left(\frac{n_k}{m_k} - \alpha\right) \geq \frac{r^+}{a\left|\frac{n_k}{m_k} - \overline{\alpha}\right|} = \nu_+ \frac{|\alpha - \overline{\alpha}|}{\left|\frac{n_k}{m_k} - \overline{\alpha}\right|}. \tag{A4.11}$$

But (A4.4) with $C \leq \dfrac{\frac{C_0}{4}}{1 + \frac{C_0}{4}}|\alpha - \overline{\alpha}|$ yields

$$\left|\frac{n_k}{m_k} - \overline{\alpha}\right| \leq \frac{1 + \frac{C_0}{2}}{1 + \frac{C_0}{4}}|\alpha - \overline{\alpha}|,$$

thus the right-hand side in (A4.11), when multiplied by $1+\frac{C_0}{2}$, yields a result not smaller than $(1+\frac{C_0}{4})\nu_+ \geq \nu'_+$.

A.5. Reminder about Borel-Laplace summation

A.5.1. General notations and properties. Let B a Banach algebra. When dealing with formal series $\sum a_n Q^n \in B[[Q]]$, it is convenient for us to use the variable $x = Q^{-1}$; we first define the *formal Borel transform* (or *formal inverse Laplace transform*) of formal series without constant term:

$$\tilde{\mathcal{L}}^{-1} : \begin{cases} x^{-1}B[[x^{-1}]] & \to & B[[\xi]] \\ \tilde{\phi} = \sum_{n\geq 0} a_n x^{-n-1} & \mapsto & \hat{\phi} = \sum_{n\geq 0} a_n \frac{\xi^n}{n!}. \end{cases}$$

Clearly, the Borel transform has nonzero radius of convergence if and only if we start with a formal Gevrey-1 series: $\tilde{\phi} \in x^{-1}B[[x^{-1}]]_1 \Leftrightarrow \tilde{\mathcal{L}}^{-1}\tilde{\phi} \in B\{\xi\}$. And starting with a convergent power-series we would obtain an entire function of exponential type in all directions.

The multiplication of Gevrey-1 formal series is tranformed into convolution of holomorphic germs:

$$\tilde{\mathcal{L}}^{-1}(\tilde{\phi}_1\tilde{\phi}_2) = \hat{\phi}_1 * \hat{\phi}_2, \qquad \hat{\phi}_i = \tilde{\mathcal{L}}^{-1}\tilde{\phi}_i, \quad \hat{\phi}_1 * \hat{\phi}_2(\xi) = \int_0^\xi \hat{\phi}_1(\xi_1)\hat{\phi}_2(\xi-\xi_1)\,d\xi_1.$$

By extending the formal Borel transform to the constant series 1, we introduce a unit δ_0 for the convolution:

$$\tilde{\mathcal{L}}^{-1} : \tilde{\phi} = \sum_{n\geq 0} a_n x^{-n} \in B[[x^{-1}]]_1 \mapsto a_0\delta_0 + \hat{\phi} \in B\delta_0 \oplus B\{\xi\}, \qquad \hat{\phi} = \sum_{n\geq 0} a_{n+1}\frac{\xi^n}{n!}.$$

We will often refer to the plane of the complex variable ξ as to the Borel plane, and to $B\{\xi\}$ or $B\delta_0 \oplus B\{\xi\}$ as to the convolutive model in contrast with the formal model $B[[x]]_1$.

The counterpart of $\partial = \frac{d}{dx}$ in the convolutive model is the multiplication by $-\xi$:

$$\tilde{\mathcal{L}}^{-1}(\partial\tilde{\phi}) = \hat{\partial}(\tilde{\mathcal{L}}^{-1}\tilde{\phi}), \qquad \hat{\partial} : \begin{cases} B\delta_0 \oplus B\{\xi\} & \to & B\{\xi\} \\ a_0\delta_0 + \hat{\phi} & \mapsto & \hat{\psi}, \end{cases} \qquad \hat{\psi}(\xi) = -\xi\hat{\phi}(\xi),$$

while multiplication by x of a series without constant term amounts essentially to differentiation with respect to ξ: if $\tilde{\phi} \in x^{-1}B[[x^{-1}]]_1$ and $\hat{\phi} = \tilde{\mathcal{L}}^{-1}\tilde{\phi}$,

$$\tilde{\mathcal{L}}^{-1}(x\tilde{\phi}) = \hat{\phi}(0)\delta_0 + \frac{d\hat{\phi}}{d\xi}.$$

A.5.2. Borel-Laplace summation. Let $\theta \in [0, 2\pi[$. Among all Gevrey-1 formal series, some of them have a Borel transform $a_0\delta_0 + \hat{\phi}$ with a holomorphic germ $\hat{\phi}$ which extends analytically along the half-line $[0, e^{i\theta}\infty[$ with at most exponential growth. In such a case one can perform the *Laplace transform of direction* θ:

$$\hat{\mathcal{L}}^\theta : a_0\delta_0 + \hat{\phi} \mapsto \phi^\theta, \qquad \phi^\theta(x) = a_0 + \int_0^{e^{i\theta}\infty} \hat{\phi}(\xi)\,e^{-x\xi}\,d\xi.$$

The resulting function ϕ^θ is holomorphic at least in a half-plane bisected by the conjugate direction (at least the half-plane $\Re e(x\, e^{i\theta}) > \delta$ if we assume $e^{-\delta|\xi|}\|\hat\phi(\xi)\|$ bounded).

If $\hat\phi$ extends analytically with at most exponential growth in a sector $\{\theta_1 \le \arg\xi \le \theta_2\}$, by moving the direction of integration and using the Cauchy Theorem we get a function analytic in a sectorial neighborhood of infinity of aperture $\pi + \theta_2 - \theta_1$. But, according to Nevanlinna's Theorem, analyticity and exponential growth in a half-strip $\{\operatorname{dist}(\xi, [0, e^{i\theta}\infty[) < \rho\}$ are sufficient to ensure that the initial formal series $\tilde\phi$ is the Gevrey-1 asymptotic expansion at infinity in a half-plane of ϕ^θ.

The interest of this process is that $\hat{\mathcal{L}}^\theta \circ \tilde{\mathcal{L}}^{-1}$ preserves multiplication, differentiation, etc., thus starting with the formal solution $\tilde\phi$ of some equation, studying the analytic continuation of $\hat\phi$ and performing $\hat{\mathcal{L}}^\theta$ for some direction θ may lead to an analytic solution of the equation (and even to distinct solutions with the same asymptotics, if analytic continuation is possible in several directions of the Borel plane with singularities in between).

A.5.3. Effect of some changes of variable. Let $\tilde\phi \in B[[x^{-1}]]_1$ and $\tilde{\mathcal{L}}^{-1}\tilde\phi = a_0\delta_0 + \hat\phi$. Let us express the formal Borel transform of $\tilde\psi(x) = \tilde\phi(f(x))$ in terms of that of $\tilde\phi$ for some elementary changes of variable f.

– For $\tilde\psi(x) = \tilde\phi(\lambda x)$ with some $\lambda \in \mathbb{C}^*$,

$$\tilde{\mathcal{L}}^{-1}\tilde\psi = a_0\delta_0 + \hat\psi, \qquad \hat\psi(\xi) = \lambda^{-1}\hat\phi(\lambda^{-1}\xi).$$

– For $\tilde\psi(x) = \tilde\phi(x+b)$ with some $b \in \mathbb{C}$,

$$\tilde{\mathcal{L}}^{-1}\tilde\psi = a_0\delta_0 + \hat\psi, \qquad \hat\psi(\xi) = e^{-b\xi}\,\hat\phi(\xi).$$

– For $\tilde\psi(x) = \tilde\phi(x + \tilde L(x))$ with some $\tilde L \in x^{-1}\mathbb{C}[[x^{-1}]]_1$ and $\hat L = \tilde{\mathcal{L}}^{-1}\tilde L$, the Taylor formula yields

$$\tilde{\mathcal{L}}^{-1}\tilde\psi = a_0\delta_0 + \hat\psi, \qquad \hat\psi = \hat\phi + \sum_{r\ge1} \hat L^{*r} * \frac{\hat\partial^r\hat\phi}{r!}, \qquad \hat L^{*r} = \underbrace{\hat L * \cdots * \hat L}_{r \text{ times}}.$$

The above series is uniformly convergent in any closed disk which is contained in the disks of convergence of $\hat\phi$ and $\hat L$. We say that $\hat\psi$ is obtained from $\hat\phi$ by *composition-convolution*, the counterpart of postcomposition by $\mathrm{Id} + L$, an operation which may look more complicated but is in fact more regularizing than postcomposition itself.

A.5.4. Simple resurgent functions. In Écalle's theory [E1], the holomorphic germ $\hat\phi$ is called the *minor* of $\tilde\phi$. The formal series $\tilde\phi$ is said to be a *simple resurgent function* if its minor satisfies the following properties:

(i) on any broken line issuing from the origin, there is a finite set of points such that $\hat\phi$ may be continued analytically along any path that closely follows the broken line in the forward direction, while circumventing (to the left or to the right) those singular points;

(ii) any determination of $\hat\phi$ in the vicinity of a singular point ω has the form

$$\hat\phi(\omega + \zeta) = \frac{c}{2\pi i\zeta} + \hat\psi(\zeta)\frac{\log\zeta}{2\pi i} + \hat R(\zeta), \qquad c \in B,\ \hat\psi, \hat R \in B\{\zeta\}.$$

A non-trivial fact is the stability under convolution of this requirement: the set of simple resurgent functions is a subalgebra of $B[[x^{-1}]]_1$. We met in Section 4.2 an example of simple resurgent function where the minor extended to a meromorphic function with simple poles only, thus a uniform (i.e. single-valued) function. But since Resurgence theory is intended to deal with nonlinear problems, and since convolution usually creates ramification, it is important that condition (i) authorise ramified and not only uniform analytic continuation.[8]

It is essential to be able to analyze the singularities which appear in the convolutive model, since they are responsible for the divergence in the formal model. This can be done by means of *alien calculus*, which relies on a family of new derivations. For each $\omega \in \mathbb{C}^*$, there is a linear operator Δ_ω of the algebra of simple resurgent functions which satisfies the Leibniz rule and measures the singular behaviour of the analytic continuation at ω of the minor of the function on which it is evaluated.

For instance, if the minor $\hat{\phi}$ is meromorphic, $\Delta_\omega\tilde{\phi} = 2\pi i \operatorname{Res}(\hat{\phi}, \omega)$. If the minor is not meromorphic but analytic on $[0, \omega[$ (the singular point ω is "viewed" from the origin, and not hidden by other singular points), $\Delta_\omega\tilde{\phi} = c + \tilde{\mathcal{L}}\hat{\psi}$ with notations as in (ii). The general formula is of the same kind but takes into account the singularities at ω of the various determinations of $\hat{\phi}$ associated to paths which follow the segment $[0, \omega[$ while circumventing the intermediary singular points.

This operator Δ_ω is called *alien derivation of index* ω by contrast with the natural derivation ∂. There is a relation

$$\Delta_\omega \circ \partial = (\partial - \omega) \circ \Delta_\omega,$$

but no relation between the alien derivations themselves: they generate a free Lie algebra. The point of view on Resurgence theory that we have indicated is rather restrictive and we refer the interested reader to [E1], [E2], [E3], [CNP] for further properties and more general definitions.

[8]For us the source of ramification was only the composition-convolution induced by some change of variable; but the fact that, when using the appropriate variable, the minor was meromorphic was related to the linear character of the problem under study.

BIBLIOGRAPHY

[ALG] J.-C. Archer and E. Le Gruyer, "On the Whitney's extension theorem," *Bull. Sci. Math.* **119** (1995), 235–266.

[Ar] V. I. Arnold, "On the mappings of the circumference onto itself," *Translations of the Amer. Math. Soc.* **46** (1961), 2nd series, 213–284.

[Be] B. Benzaghou, "Algèbres de Hadamard," *Bull. de la Soc. Math. de France* **98** (1970), 209–252.

[BMS] A. Berretti, S. Marmi and D. Sauzin, "Limit at resonances of linearizations of some complex analytic dynamical systems," *Ergodic Theory and Dynamical Systems* **20** (2000), 963–990.

[Be1] A. Beurling, "Sur les fonctions limites quasi-analytiques des fractions rationelles," *8th Scandinavian Math. Congress*, Stockholm (1934), 199–210; in *Collected Works of Arne Beurling*, Birkhäuser, Boston Basel Berlin, Vol. I (1989), 109–120.

[Be2] A. Beurling, "On quasianalyticity and general distributions," *Multilithed Lecture Notes*, Summer School, Stanford University (1961); in *Collected Works of Arne Beurling*, Birkhäuser, Boston Basel Berlin, Vol. I (1989), 309–338.

[Bo] E. Borel, *Leçons sur les fonctions monogènes uniformes d'une variable complexe*, Gauthier-Villars, Paris (1917).

[CNP] B. Candelpergher, J.-C. Nosmas and F. Pham, *Approche de la résurgence*, Actualités Math., Hermann, Paris (1993).

[CCD] B. Candelpergher, M.-A. Coppo and E. Delabaere, "La sommation de Ramanujan," *L'Enseignement mathématique* **43** (1997), 93–132.

[Ca] T. Carleman, *Les fonctions quasi analytiques*, Gauthier-Villars, Paris (1926).

[De] A. Denjoy, "Sur les séries de fractions rationelles," *Bull. de la Soc. Math. de France* **52** (1924), 418–434.

[Du] D. Duverney, "Explicit computation of Padé-Hermite approximants," *J. Approx. Th.* **88** (1997), 80–91.

[E1] J. Écalle, *Les fonctions résurgentes et leurs applications*, Publ. math. d'Orsay, vol. I: **81–05**, vol. II: **81–06**, vol. III: **85–05** (1981, 1985).

[E2] J. Écalle, "Singularités non abordables par la géométrie," *Ann. Inst. Fourier, Grenoble* **42** (1992), 73–194.

[E3] J. Écalle, *Introduction aux fonctions analysables et preuve constructive de la conjecture de Dulac*, Hermann, Paris (1992).

[Ga] T. W. Gamelin, *Uniform Algebras*, Prentice-Hall, Engelwood Cliffs (1969).

[Gam] J. L. Gammel, "Continuation of functions beyond natural boundaries," *Rocky Mountain J. Math.* **4** (1974), 203–206.

[GN] J. L. Gammel and J. Nuttall, "Convergence of Padé Approximants to quasianalytic functions beyond natural boundaries," *J. Math. Analys. Appl.* **43** (1973), 694–696.

[Gl] G. Glaeser, "Étude de quelques algèbres tayloriennes," *J. Anal. Math. Jerusalem* **6** (1958), 1–124.

[Gou] E. Goursat, "Sur les fonctions à espaces lacunaires," *Bull. Sci. Math.* **11** (1887), 109–114.

[HL] G. H. Hardy and J. E. Littlewood, "Notes on the theory of series (XXIV): a curious power-series," *Proc. Cambridge Phil. Soc.* **42** (1946), 85–88.

[HW] G. H. Hardy and E. M. Wright, *An introduction to the theory of numbers*, Oxford University Press (1979).

[He] M. R. Herman, "Simple proofs of local conjugacy theorems for diffeomorphisms of the circle with almost every rotation numbers," *Bull. Soc. Bras. Mat.* **16** (1985), 45–83.

[Khi] A. Ya. Khinchin, *Continued Fractions*, The University of Chicago Press, Chicago London (1964).

[Ko] A. N. Kolmogorov, "The General Theory of Dynamical Systems and Classical Mechanics", address to the 1954 International Congress of Mathematicians, Amsterdam.

[La] S. Lang, *Elliptic functions*, Graduate Texts in Math. **112**, Springer-Verlag (1987).

[Ma] B. Malgrange, "Sommation des séries divergentes," *Expos. Math.* **13** (1995), 163–222.

[MMY] S. Marmi, P. Moussa and J.-C. Yoccoz, "The Brjuno functions and their regularity properties," *Comm. Math. Phys.* **186** (1997), 265–293.

[Mü] J. Müller, "The Hadamard Multiplication Theorem and Applications in Summability Theory," *Complex Variables* **18** (1992), 155-166.

[P1] H. Poincaré, "Sur les fonctions à espaces lacunaires," *Amer. J. Math.* **14** (1892), 201–221.

[P2] H. Poincaré, "Analyse de ses travaux sur la théorie générale des fonctions d'une variable," *Acta Math.* **38** (1921), 65–70.

[Pö] J. Pöschel, "Integrability of Hamiltonian systems on Cantor sets," *Comm. Pure Appl. Math.* **35** (1982), 653–696.

[Ra] J.-P. Ramis, *Séries divergentes et théories asymptotiques*, Panoramas et synthèses, Suppl. au Bull. de la Soc. Math. de France **121** (1993).

[Re] R. Remmert, *Classical Topics in Complex Function Theory*, Graduate Texts in Math. **172**, Springer-Verlag (1998).

[Ris] E. Risler, *Linéarisation des perturbations holomorphes des rotations et applications*, Mémoires de la Soc. Math. de France **77** (1999).

[Sc] S. Schottlaender, "Der Hadamardsche Multiplicationssatz und weitere Kompositions sätze der Funktionentheorie," *Math. Nachr.* **11** (1954), 239–294.

[Si] R. V. Sibiliev, "Uniqueness theorems for Wolff-Denjoy series," *St. Petersburg Math. J.* **7** (1996), 145–168.

[Sim] B. Simon, "Almost periodic Schrödinger operators IV. The Maryland model," *Ann. Phys.* **159** (1985), 157–183.

[St] E. M. Stein, *Singular integrals and differentiability properties of functions*, Princeton Math. Series **30** (1970).

[Th] V. Thilliez, "Quelques propriétés de quasi-analyticité," *Gazette des Mathematiciens* **70** (1996), 49–68.

[Tr] F. G. Tricomi, "Determinazione del valore di un classico prodotto infinito," *Rend. Acc. Naz. Lincei Classe Sci. Fis. Mat. Nat.* **XIV** (1953), 3–7.

[Tj] W. J. Trjitzinsky, "On Quasi-Analytic Functions," *Ann. of Math.* **30** (1930), 526–546.

[We] A. Weil, *Elliptic Functions according to Eisenstein and Kronecker*, Springer-Verlag, Berlin Heidelberg New York (1976).

[Wh] H. Whitney, "Analytic extensions of differentiable functions defined in closed sets," *Trans. Amer. Math. Soc.* **36** (1934), 63–89.

[Wk] J. Winkler, "A uniqueness theorem for monogenic functions," *Ann. Acad. Sci. Fennicae Ser. A. I. Math.* **18** (1993), 105–116.

[Wi] A. Wintner, "The linear difference equation of first order for angular variables," *Duke Math. J.* **12** (1945), 445–449.

[Wo] J. Wolff, "Sur les séries $\sum \frac{A_n}{z-z_n}$," *Comptes Rendus Acad. Sci. Paris* **173** (1921), 1327–1328.

[Y1] J.-C. Yoccoz, "Conjugaison des difféomorphismes analytiques du cercle," manuscript (1988).

[Y2] J.-C. Yoccoz, "Petits diviseurs en dimension un," *Astérisque* **231** (1995).

[Y3] J.-C. Yoccoz, "Analytic linearization of analytic circle diffeomorphisms," lectures given at the CIME school "Dynamical Systems and Small Divisors", Cetraro (Italy) 1998, to appear in the *CIME series of Lecture Notes in Math.*

[Za] L. Zalcman, *Analytic capacity and rational approximation*, Lecture Notes in Math. **50**, Springer (1968).

Editorial Information

To be published in the *Memoirs*, a paper must be correct, new, nontrivial, and significant. Further, it must be well written and of interest to a substantial number of mathematicians. Piecemeal results, such as an inconclusive step toward an unproved major theorem or a minor variation on a known result, are in general not acceptable for publication. Papers appearing in *Memoirs* are generally longer than those appearing in *Transactions*, which shares the same editorial committee.

As of April 1, 2003, the backlog for this journal was approximately 4 volumes. This estimate is the result of dividing the number of manuscripts for this journal in the Providence office that have not yet gone to the printer on the above date by the average number of monographs per volume over the previous twelve months, reduced by the number of volumes published in four months (the time necessary for preparing a volume for the printer). (There are 6 volumes per year, each containing at least 4 numbers.)

A Consent to Publish and Copyright Agreement is required before a paper will be published in the *Memoirs*. After a paper is accepted for publication, the Providence office will send a Consent to Publish and Copyright Agreement to all authors of the paper. By submitting a paper to the *Memoirs*, authors certify that the results have not been submitted to nor are they under consideration for publication by another journal, conference proceedings, or similar publication.

Information for Authors

Memoirs are printed from camera copy fully prepared by the author. This means that the finished book will look exactly like the copy submitted.

The paper must contain a *descriptive title* and an *abstract* that summarizes the article in language suitable for workers in the general field (algebra, analysis, etc.). The *descriptive title* should be short, but informative; useless or vague phrases such as "some remarks about" or "concerning" should be avoided. The *abstract* should be at least one complete sentence, and at most 300 words. Included with the footnotes to the paper should be the 2000 *Mathematics Subject Classification* representing the primary and secondary subjects of the article. The classifications are accessible from `www.ams.org/msc/`. The list of classifications is also available in print starting with the 1999 annual index of *Mathematical Reviews*. The Mathematics Subject Classification footnote may be followed by a list of *key words and phrases* describing the subject matter of the article and taken from it. Journal abbreviations used in bibliographies are listed in the latest *Mathematical Reviews* annual index. The series abbreviations are also accessible from `www.ams.org/publications/`. To help in preparing and verifying references, the AMS offers MR Lookup, a Reference Tool for Linking, at `www.ams.org/mrlookup/`. When the manuscript is submitted, authors should supply the editor with electronic addresses if available. These will be printed after the postal address at the end of the article.

Electronically prepared manuscripts. The AMS encourages electronically prepared manuscripts, with a strong preference for $\mathcal{A}_{\mathcal{M}}\mathcal{S}$-LaTeX. To this end, the Society has prepared $\mathcal{A}_{\mathcal{M}}\mathcal{S}$-LaTeX author packages for each AMS publication. Author packages include instructions for preparing electronic manuscripts, the *AMS Author Handbook*, samples, and a style file that generates the particular design specifications of that publication series. Though $\mathcal{A}_{\mathcal{M}}\mathcal{S}$-LaTeX is the highly preferred format of TeX, author packages are also available in $\mathcal{A}_{\mathcal{M}}\mathcal{S}$-TeX.

Authors may retrieve an author package from e-MATH starting from www.ams.org/tex/ or via FTP to ftp.ams.org (login as anonymous, enter username as password, and type cd pub/author-info). The *AMS Author Handbook* and the *Instruction Manual* are available in PDF format following the author packages link from www.ams.org/tex/. The author package can be obtained free of charge by sending email to pub@ams.org (Internet) or from the Publication Division, American Mathematical Society, 201 Charles St., Providence, RI 02904, USA. When requesting an author package, please specify AMS-LaTeX or AMS-TeX, Macintosh or IBM (3.5) format, and the publication in which your paper will appear. Please be sure to include your complete mailing address.

Sending electronic files. After acceptance, the source file(s) should be sent to the Providence office (this includes any TeX source file, any graphics files, and the DVI or PostScript file).

Before sending the source file, be sure you have proofread your paper carefully. The files you send must be the EXACT files used to generate the proof copy that was accepted for publication. For all publications, authors are required to send a printed copy of their paper, which exactly matches the copy approved for publication, along with any graphics that will appear in the paper.

TeX files may be submitted by email, FTP, or on diskette. The DVI file(s) and PostScript files should be submitted only by FTP or on diskette unless they are encoded properly to submit through email. (DVI files are binary and PostScript files tend to be very large.)

Electronically prepared manuscripts can be sent via email to pub-submit@ams.org (Internet). The subject line of the message should include the publication code to identify it as a Memoir. TeX source files, DVI files, and PostScript files can be transferred over the Internet by FTP to the Internet node e-math.ams.org (130.44.1.100).

Electronic graphics. Comprehensive instructions on preparing graphics are available at www.ams.org/jourhtml/graphics.html. A few of the major requirements are given here.

Submit files for graphics as EPS (Encapsulated PostScript) files. This includes graphics originated via a graphics application as well as scanned photographs or other computer-generated images. If this is not possible, TIFF files are acceptable as long as they can be opened in Adobe Photoshop or Illustrator. No matter what method was used to produce the graphic, it is necessary to provide a paper copy to the AMS.

Authors using graphics packages for the creation of electronic art should also avoid the use of any lines thinner than 0.5 points in width. Many graphics packages allow the user to specify a "hairline" for a very thin line. Hairlines often look acceptable when proofed on a typical laser printer. However, when produced on a high-resolution laser imagesetter, hairlines become nearly invisible and will be lost entirely in the final printing process.

Screens should be set to values between 15% and 85%. Screens which fall outside of this range are too light or too dark to print correctly. Variations of screens within a graphic should be no less than 10%.

Inquiries. Any inquiries concerning a paper that has been accepted for publication should be sent directly to the Electronic Prepress Department, American Mathematical Society, 201 Charles St., Providence, RI 02904, USA.

Editors

This journal is designed particularly for long research papers, normally at least 80 pages in length, and groups of cognate papers in pure and applied mathematics. Papers intended for publication in the *Memoirs* should be addressed to one of the following editors. In principle the Memoirs welcomes electronic submissions, and some of the editors, those whose names appear below with an asterisk (*), have indicated that they prefer them. However, editors reserve the right to request hard copies after papers have been submitted electronically. Authors are advised to make preliminary email inquiries to editors about whether they are likely to be able to handle submissions in a particular electronic form.

Algebra to KAREN E. SMITH, Department of Mathematics, University of Michigan, 525 University, Suite 2832, Ann Arbor, MI 48109-1109; email: kesmith@lsa.umich.edu

Algebraic geometry and commutative algebra to LAWRENCE EIN, Department of Mathematics, University of Illinois, 851 S. Morgan (M/C 249), Chicago, IL 60607-7045; email: ein@uic.edu

Algebraic topology and cohomology of groups to STEWART PRIDDY, Department of Mathematics, Northwestern University, 2033 Sheridan Road, Evanston, IL 60208-2730; email: priddy@math.nwu.edu

Combinatorics and Lie theory to SERGEY FOMIN, Department of Mathematics, University of Michigan, Ann Arbor, Michigan 48109-1109; email: fomin@umich.edu

Complex analysis and complex geometry to DUONG H. PHONG, Department of Mathematics, Columbia University, 2990 Broadway, New York, NY 10027-0029; email: phong@math.columbia.edu

*__Differential geometry and global analysis__ to LISA C. JEFFREY, Department of Mathematics, University of Toronto, 100 St. George St., Toronto, ON Canada M5S 3G3; email: jeffrey@math.toronto.edu

Dynamical systems and ergodic theory to ROBERT F. WILLIAMS, Department of Mathematics, University of Texas, Austin, Texas 78712-1082; email: bob@math.utexas.edu

Functional analysis and operator algebras to DAN VOICULESCU, Department of Mathematics, University of California, Berkeley, 970 Evans Hall, Floor 9, Berkeley, CA 94720-0001; email: dvv@math.berkeley.edu

Geometric topology, knot theory and hyperbolic geometry to ABIGAIL A. THOMPSON, Department of Mathematics, University of California, Davis, Davis, CA 95616-5224; email: thompson@math.ucdavis.edu

Harmonic analysis to ALEXANDER NAGEL, Department of Mathematics, University of Wisconsin, 480 Lincoln Drive, Madison, WI 53706-1313; email: nagel@math.wisc.edu

Harmonic analysis, representation theory, and Lie theory to ROBERT J. STANTON, Department of Mathematics, The Ohio State University, 231 West 18th Avenue, Columbus, OH 43210-1174; email: stanton@math.ohio-state.edu

*__Logic__ to THEODORE SLAMAN, Department of Mathematics, University of California, Berkeley, CA 94720-3840; email: slaman@math.berkeley.edu

Number theory to HAROLD G. DIAMOND, Department of Mathematics, University of Illinois, 1409 W. Green St., Urbana, IL 61801-2917; email: diamond@math.uiuc.edu

*__Ordinary differential equations, and applied mathematics__ to PETER W. BATES, Department of Mathematics, Michigan State University, East Lansing, MI 48824-1027; email: peter@math.msu.edu

*__Partial differential equations__ to PATRICIA E. BAUMAN, Department of Mathematics, Purdue University, West Lafayette, IN 47907-1395' email: bauman@math.purdue.edu

*__Probability and statistics__ to KRZYSZTOF BURDZY, Department of Mathematics, University of Washington, Box 354350, Seattle, Washington 98195-4350; email: burdzy@math.washington.edu

*__Real analysis and partial differential equations__ to DANIEL TATARU, Department of Mathematics, University of California, Berkeley, Berkeley, CA 94720; email: tataru@math.berkeley.edu

All other communications to the editors should be addressed to the Managing Editor, WILLIAM BECKNER, Department of Mathematics, University of Texas, Austin, TX 78712-1082; email: beckner@math.utexas.edu.

Titles in This Series

783 **Ethan Akin, Mike Hurley, and Judy A. Kennedy,** Dynamics of topologically generic homeomorphisms, 2003

782 **Masaaki Furusawa and Joseph A. Shalika,** On central critical values of the degree four L-functions for GSp(4): The Fundamental Lemma, 2003

781 **Marcin Bownik,** Anisotropic Hardy spaces and wavelets, 2003

780 **S. Marmi and D. Sauzin,** Quasianalytic monogenic solutions of a cohomological equation, 2003

779 **Hansjörg Geiges,** h-principles and flexibility in geometry, 2003

778 **David B. Massey,** Numerical control over complex analytic singularities, 2003

777 **Robert Lauter,** Pseudodifferential analysis on conformally compact spaces, 2003

776 **U. Haagerup, H. P. Rosenthal, and F. A. Sukochev,** Banach embedding properties of non-commutative L^p-spaces, 2003

775 **P. Lochak, J.-P. Marco, and D. Sauzin,** On the splitting of invariant manifolds in multidimensional near-integrable Hamiltonian systems, 2003

774 **Kai A. Behrend,** Derived ℓ-adic categories for algebraic stacks, 2003

773 **Robert M. Guralnick, Peter Müller, and Jan Saxl,** The rational function analogue of a question of Schur and exceptionality of permutation representations, 2003

772 **Katrina Barron,** The moduli space of $N = 1$ superspheres with tubes and the sewing operation, 2003

771 **Shigenori Matsumoto,** Affine flows on 3-manifolds, 2003

770 **W. N. Everitt and L. Markus,** Elliptic partial differential operators and symplectic algebra, 2003

769 **Jie Wu,** Homotopy theory of the suspensions of the projective plane, 2003

768 **R. Höpfner and E. Löcherbach,** Limit theorems for null recurrent Markov processes, 2003

767 **Po Hu,** S-modules in the category of schemes, 2003

766 **Su Gao and Alexander S. Kechris,** On the classification of Polish metric spaces up to isometry, 2003

765 **Robert Bieri and Ross Geoghegan,** Connectivity properties of group actions on non-positively curved spaces, 2003

764 **J. Spandaw,** Noether-Lefschetz problems for degeneracy loci, 2003

763 **Yasuyuki Kachi and Eiichi Sato,** Segre's reflexivity and an inductive characterization os hyperquadrics, 2002

762 **Leiba Rodman, Ilya M. Spitkovsky, and Hugo Woerdeman,** Abstract band method via factorization, positive and band extensions of multivariable almost periodic matrix functions, and spectral estimation, 2002

761 **Oliver Druet and Emmanuel Hebey,** The AB program in geometric analysis : Sharp Sobolev inequalities and related problems, 2002

760 **Markus Banagl,** Extending intersection homology type invarients to non-Witt spaces, 2002

759 **Donald M. Davis,** From representation theory to homotopy groups, 2002

758 **Alan Forrest, John Hunton, and Johannes Kellendonk,** Topological invariants for projection method patterns, 2002

757 **Douglas Bowman,** q-difference operators, orthogonal polynomials, and symmetric expansions, 2002

756 **José Ignacio Cogolludo-Agustín,** Topological invariants of the complement to arrangements of rational plane curves, 2002

755 **M. A. Mandell and J. P. May,** Equivariant orthogonal spectra and S-modules, 2002

[redacted]

[redacted]

For a complete list of titles in this series, visit the
AMS Bookstore at **www.ams.org/bookstore/**.

[redacted]